T0155838

SpringerBriefs in Applied Sciences and Technology

PoliMI SpringerBriefs

More information about this subseries at http://www.springer.com/series/11159

http://www.polimi.it

Marcella M. Bonanomi

Digital Transformation of Multidisciplinary Design Firms

A Systematic Analysis-Based Methodology for Organizational Change Management

POLITECNICO
MILANO 1863

Marcella M. Bonanomi
Innovative and Industrial Construction
ETH Zurich
Zürich, Switzerland

ISSN 2191-530X   ISSN 2191-5318   (electronic)
SpringerBriefs in Applied Sciences and Technology
ISSN 2282-2577   ISSN 2282-2585   (electronic)
PoliMI SpringerBriefs
ISBN 978-3-030-19700-1   ISBN 978-3-030-19701-8   (eBook)
https://doi.org/10.1007/978-3-030-19701-8

This Springer imprint is published by the registered company Springer Nature Switzerland AG.
The registered company address is: Gewerbestrasse 11, 6330 Cham, Switzerland

Preface

The work presented in this publication is divided into six chapters organized into three parts. The first part introduces the theoretical background.

Chapter 1 aims to underline the opportunity of innovating the Architecture, Engineering, and Construction (AEC) industry by embracing digital processes and technologies. Specifically, the chapter outlines digital transformation trends that the AEC industry is witnessing and the contribution that they can offer in supporting firms to embrace new forms of process and organization.

Chapter 2 aims to present the digital transformation strategies of multidisciplinary design firms, which represent the investigation field of this work. To this purpose, experts' interviews have been performed and the data collected are framed according to four lenses of interpretation: process, people, product, and policy. The combination of the literature review and the experts' interviews leads to identify the paradigm shifts of the digital transformation of multidisciplinary design firms. These paradigm shifts represent the point of departure of this study: from a silo-based and sequential to a collaborative and iterative digital design process; from a vertical and hierarchical to a platform-based and networked organization.

Starting from these paradigm shifts, Chap. 3 aims to propose a theoretical framework to manage the digital transformation of multidisciplinary design firms. This framework is built on the combination of three conceptual constructs: paradigms, perspectives, and context of change. The proposed paradigms of change are collaboration, iteration, platform, and network. Process and organization represent the two diverse, but interacting, perspectives of change for digital transformation, while the context of change must be understood as the combination of reference market, institutions, and firm's specificity.

The second part introduces a methodological experimentation.

Chapter 4 aims to propose a systematic methodology for organizational change management to identify, assess, and evaluate the process-oriented and organizational changes, which occur in multidisciplinary design firms in periods of digital transformation. Specifically, the methodology proposed consists of a framework to identify, categorize, and structure these changes and a set of analysis methods to assess, visualize, and evaluate them. The set of change analysis methods proposed includes process mapping, stakeholder mapping, timesheet tracking, and Social Network Analysis (SNA).

In the third part, the results of the application are presented and discussed.

Chapters 5 and 6 apply and test the systematic methodology proposed in two real-world case studies. Specifically, Chap. 5 aims to apply and test the methodology in an Italian multidisciplinary design firm to support the firm in transitioning to a collaborative and iterative digital design process. Chapter 6 aims to apply and test the proposed methodology in a Canadian multidisciplinary design firm to support the firm in transitioning to a platform-based and networked organizational structure. The two case studies apply and test separately the two perspectives of change for digital transformation proposed by the methodology developed. Specifically, the process-oriented perspective is applied and tested in the Italian case study, while the organizational one in the Canadian case study. The main results are presented and commented upon.

Zürich, Switzerland   Dr. Marcella M. Bonanomi

Acknowledgements

Foremost, I would like to express my sincere gratitude to my advisor, Prof. Cinzia Talamo, for her guidance and support. She constantly shared her knowledge, intuitions, and passion during my Ph.D. For the same reason, I would also like to acknowledge Prof. Giancarlo Paganin and Prof. Andrea Campioli, co-advisors with Prof. Talamo of the work.

Sincere thanks go to firm X, and his Chief in particular, for the enthusiastic cooperation, and for providing a case study to run applications. With him, I would also like to thank the entire staff. Many thanks also to Nazly Atta and Marco Migliore for being good friends before than colleagues.

With reference to the Canadian experience, my gratitude goes to Prof. Sheryl Staub-French for the warm welcome she gave me at the BIM TOPiCS Lab at The University of British Columbia. She was the busiest and most available person at the same time, being always able to give me valuable suggestions for my work and career. A very special thanks also to Aubrey Tucker. It was such a pleasure working with him and learning from him.

Finally, none of this work would have been possible without the love and patience of all my family and friends. In particular, I thank my wonderful parents, source of unconditional love and support and from whom I got my strength and determination, and the capability to dream big. I thank my sister and brother for believing in me when I doubted about myself and for encouraging me to always set higher goals and put the greatest effort to achieve them. I owe them everything. Last but not least, I thank my dog Wilson for always being next to me in the most gloomy and rainy days remembering me that the time for being happy is now.

Acknowledgements

Introduction

The digitalization of the Architecture, Engineering, and Construction (AEC) industry has the potential to transform the process through which buildings are designed and built, as well as the relationships through which practitioners work and interact. Certain activities of the traditional building process disappear, while others appear, distribution of work is reviewed, and new relationships, roles, and responsibilities emerge within firms.

With reference to the significant impact of digitalization on processes and organizations, many players increasingly recognize that the technological aspects of this digital transformation are the 'easiest' to manage; the more significant challenges relate to the process-oriented and organizational changes.

Few studies have investigated the topic of organizational change management in the specific context of the digitalization of the Architecture and Engineering (A/E) firms. As such, there are few approaches and methods to study the process-oriented and organizational changes associated with the digital transformation of multidisciplinary design firms. However, some work has been performed to investigate the implementation of digital technologies and processes, such as Building Information Modeling (BIM), across the industry in general, and in architectural practices in particular. Additionally, over the past three decades, many studies have been done about the context of innovation, its process, and its outcomes.

Given these considerations, the present study aims to bridge the gap between the growing literature about digital transformation and the established literature on innovation by developing new knowledge constructs and proposing analysis methods to characterize organizational change management in the context of the digitalization of multidisciplinary design firms.

Starting from this point of departure, the general objective of this study is to investigate how digitalization effects processes and organizational structures of multidisciplinary design firms. To achieve this overall objective, this work focuses on three different, but interacting levels (specific objectives):

1. To define a theoretical framework to identify, structure, and categorize the changes brought on by the digital transformation within multidisciplinary design firms;
2. To propose a set of qualitative and quantitative methods to identify, visualize, and assess these changes;
3. To develop a systematic methodology for organizational change management, which includes both the aforementioned framework and set of change analysis methods.

Contents

**PART III Applications: Test of the Methodology Proposed
on Case Studies**

About the Author

Dr. Marcella M. Bonanomi is Postdoctoral Research Associate at the Chair of Innovative and Industrial Construction (IC) of the Institute of Construction and Infrastructure Management (IBI) at the ETH Zürich since September 2018. The research interest of her is the digital transformation of the AEC industry, and the related impact on processes and organizations. Specifically, her work focuses on enhancing digital innovation in AEC firms through a transformation from silo-based processes and static-hierarchical organizations to collaborative digital workflows and platform-based and networks of teams. Marcella holds a Doctor of Philosophy (2018) in architecture, built environment and construction engineering (ABC) from the Politecnico di Milano (Italy). Her dissertation investigated the digital transformation of multidisciplinary design firms and the related process-oriented and organizational changes. In 2017, Marcella has been Visiting Ph.D. Scholar at the BIM TOPiCS Lab at The University of British Columbia (UBC) in Vancouver (Canada). While at UBC, she was awarded the Visiting International Research Students (VIRS) Grant from The University of British Columbia. In addition, she was a member of the Innovative Technology Development Team at Stantec, an international leading AEC firm, and of the R&D BIM team of a large Italian multidisciplinary design firm. From 2013 to 2014, she also worked as Research Fellow at the Politecnico di Milano, where she investigated the use of building information modeling (BIM) to support facilities management (FM) activities. In the last five years, she has assisted courses at the Politecnico di Milano and ETH Zurich, presented her work at national and international conferences, and published several articles about BIM implementation, digital transformation, and the related organizational change management. She has also co-authored two books about knowledge management and sustainability, as well as wrote blog posts and white papers on key business topics. She holds a Master of Science (2013) in architecture from the Politecnico di Milano and Politecnico di Torino and a Bachelor of Science (2010) in architectural design from the Politecnico di Milano. In the last year of her M.Sc. studies, she did a six-month exchange period at the Staatliche Akademie der Bildenden Künste Stuttgart in Germany. From 2010 to 2013, she has been doing internships in leading architecture and interior design firms in Milan, such as Lissoni Associati and Metrogramma. She is also an alumnus of the Alta Scuola Politecnica (ASP), an excellence master school with a competitive admission process and extracurricular activities focused on innovative and multidisciplinary applied research projects.

Part I
Theoretical Background: Process-Oriented and Organizational Changes Associated with the Digital Transformation of Multidisciplinary Design Firms

Chapter 1
Digital Transformation Trends: Toward New Forms of Process and Organization

Abstract The aim of this chapter is to present the digital transformation trends that the Architecture, Engineering, and Construction (AEC) industry is witnessing. The purpose of this review is to illustrate the changes brought on by the digitalization specifically within architectural and engineering firms from two perspectives: process and organization. To this purpose, the following trends are presented to describe the changes that the traditional design process and firms' organizational structures respectively undergo in periods of digital transformation: (1) Design with data for empowering 3D with D^3 (Data-Driven-Decisions), and (2) Platform-based and networked organizations for leveraging the power of ecosystems. Lastly, a third trend is introduced to outline a new form of project stakeholder management through a network approach.

Keywords Digital transformation · Architecture and engineering · Design process · Data · Platform · Network · Organizational change

1.1 Design with Data: Empowering 3D with D^3 (Data-Driven Decisions)

Optimization of building performance and quality is one of the challenge associated with the digitalization of the Architecture, Engineering, and Construction (AEC) industry. To this purpose, coupling 3D models with D^3 (Data-Driven Decisions) is a new way of working that digitally-mature architectural and engineering firms have already started to implement (Deutsch 2015). The objective of this new approach to designing is to enable a data-informed design input to lead toward a performance-based design output and therefore to ensure a better building quality (Haponava and Al-Jibouri 2010). About this matter of data informing design outcomes, Tassera (2017), computational BIM Manager at Ridley,[1] has recently referred a statement

---

[1] Ridley is a design and construction firm which has offices across major cities in Australia, Singapore, Hong Kong and North America. The firm is seeking to implement and use building data to drive change within the industry towards a data-informed design outcome. They are also helping their clients to recognize the value of building data management.

3

made by one of the business clients: "Buildings equal data". This statement points out the firm's vision and mission, which is the idea of an "Intelligent Built Environment" shaped on the convergence of physical and digital worlds.

Given these considerations, architectural and engineering firms have started to simulate design solutions through the means of data in order to evaluate project alternatives and achieve the optimal project performance according to a set of relevant parameters (Zanchetta et al. 2014). Simulation and iteration are increasingly becoming the lenses through which to assess design process innovation over the long term.

Simulation and iteration are strictly related to the concept of collaboration. On the contrary, the traditional design process is usually organized as a sequence of silo-based tasks in which stakeholders from different disciplines work autonomously according to a consequential decision making (Zhongbao and Xiangfeng 2013). To make an example, structural engineers usually develop their project starting from a pre-defined architectural design. In other words, the structural project is most of the times not collaboratively and iteratively designed together with the architectural team but rather adapted ex-post to the architectural project. This lack of collaboration and knowledge sharing between the project participants negatively affects both the design outcome, in terms of building quality, and the design process, in terms of cost and time (Lindhard and Larsen 2016).

Based on this, concepts, such as data-informed, performance-based, and service-life planning, are believed to be the attributes of a new 'theory' of the design process. The idea of promoting building quality as a function of performance over time[2] is set to become a significant design factor (Chong et al. 2014; Walbe Ornstein et al. 2009). The possibility provided by digital technologies and processes, such as Building Information Modelling (BIM), to create collaboratively and iteratively data-enriched 3D models can help to move towards this direction. The strategy should be to increasingly employ a simulative, iterative, and collaborative approach in order to determine a positive impact on building quality, process efficiency, and sector productivity.

1.2 Platform-Based and Networked Organizations: Leveraging the Power of Ecosystems

In a recent article published in the Harward Business Review, Schrage (2016) outlines an emergent trend that he defines as the "Platformization" of organizational structures. As Schrage (2016) states: "The ongoing digitalization and 'networkification' of business processes render organizational structures uniquely suitable to be

---

[2]This trend is also justified by the fact that the R&M (Renovation and Maintenance) market is experiencing a significant growth. Therefore, durability issues play a crucial role in terms of building quality, as well as cost savings and time efficiency. Durability management, as well as Estimated Service Life (ESL), allows not only to improve and optimize preventive maintenance of built environment, but also to carry out Life-Cycle Assessments (LCA) and Life-Cycle Costing (LCC) analyses.

platformed." A similar viewpoint is also endorsed by the BCG report entitled "The Transformative Power of BIM" (Gerbert et al. 2016) in which the following organizational change strategies are proposed in order to successfully manage a firm's digital transformation:

– To reconfigure organizational structures in response to the new paradigms of digital innovation: platform and network. Digital technologies and processes require collaborative interactions, rather than silo-based transactions, across process tasks and between project parties;
– To define new roles and responsibilities by either hiring or training people at both strategic and operational levels. The definition of new roles and responsibilities is strictly associated with organizational structure change. As a recent article on the McKinsey Digital blog argues, "[…] adding these new roles without an aligned operating model can actually lead to more confusion, power struggles, and a negative effect on the firm's overall performance" (Eizenman et al. 2017);
– To complement internal digital knowledge through third parties by networking with suppliers and clients to strengthen the overall supply chain. Firms in digital transition should not only reconfigure their organizational structures, but also to foster a change in project stakeholder management (see Sect. 1.3).

The shift toward platform-based and networked structures (Picon 2016) matches and reflects the same mechanism that underlies a new knowledge management (KM) approach which is increasingly emerging within digitally-mature design firms. As design firms start implementing digital technologies and processes, their organizational structures must be reconfigured not only to align with new roles and responsibilities (Gerbert et al. 2016) but also to support an efficient and effective knowledge management.[3] Many studies have focused on how organizational structures can affect the diffusion of information (Allen 1977; Burt 1992; Rogers 1995). With reference to a 'new' organizational knowledge management, Hegazy and Ghorab (2014) argue that "The new paradigm is that within the organization knowledge must be shared for it to grow".

To conclude, the emergence of these new forms of organization, which operate according to a service-based approach and facilitating knowledge sharing, is a strong argument to highlight the need for more project stakeholders' integration, above all end-users and clients (Al Ahbabi and Alshawi 2015).

---

[3]Duhon (1998) defines knowledge management as "A discipline that promotes an integrated approach to identifying, capturing, evaluating, retrieving, and sharing all of an enterprise's information assets".

1.3 Network Approach: A Revolution in Project Stakeholder Management

The implementation of digital technologies and processes requires architectural and engineering firms to rethink their traditional supply chain management (SCM) (Papadonikolaki et al. 2016). This also leads to changes in the traditional approach to knowledge management and risk management (Eastman 2011). Insights about changes brought on by the digitalization in these three categories are proposed below.

1.3.1 Supply Chain Management

The conventional approach according to which project stakeholders work in silos focusing mainly on their own incentives and profits does not answer the increasing need for an integrated approach to project design and delivery (Deutsch 2015). Frequent results of this traditional approach include:

– Data inconsistency/redundancy;
– Process inefficiencies;
– Project underperformance;
– Adversarial relationships (Czmocha and Pekala 2014).

Given the importance of project stakeholders' integration, design firms are increasingly asked to reconfigure their traditional supply chain relationships into digital supply network partnerships (Mussomeli et al. 2016; Papadonikolaki and Wamelink 2017; Vasudevan et al. 2018). Accordingly, the Accenture report (Hanifan et al. 2014) entitled "The Digital Supply Network: A New Paradigm for Supply Chain Management" states: "In a metaphorical sense, people and data—as well as materials, products, and supplies—must travel together across the extended enterprise." Based on this, the shift from supply chains to supply networks affects people behaviors and data streams, and therefore knowledge management.

1.3.2 Knowledge Management

A recent report by McKinsey (Eizenman et al. 2017), entitled "Reinventing Construction: A Route to Higher Productivity", suggests three major interventions to enhance knowledge management:

– Embedding innovation throughout the organization and across the supply chain;
– Strengthening the links with the supply chain, above all with technology suppliers and owners;
– Improving risk sharing of new approaches to project design and delivery.

Given these considerations, design firms should move away from traditional silo-based knowledge management and move toward a collaborative approach fostering project stakeholders' integration. Not only design firms, but rather all project parties, should embrace this change. Technology providers, for example, must meet the growing needs of design firms by (1) developing collaborative and integrated technology ecosystems, (2) strengthening the interoperability of their products, and (3) striving for open standards to address critical interfaces between software and tools.

Other project stakeholders must make appropriate adjustments to facilitate and support better knowledge creation and distribution. Owners in particular are increasingly asked to play a key role in this regard (Al Ahbabi and Alshawi 2015; McGraw Hill 2014). However, despite positive impacts and big business value of digital technologies and processes for owners, a survey conducted by McGraw Hill (2014) shows, for example, that just 20% of the UK building owners report having and using BIM authoring software. These data underline two significant challenges: (1) Clients are still not enough involved in the process, and (2) Clients are not conscious of the primary role they should play (Succar 2010). Based on this, there seems to be scope for a new project management and project delivery by fostering project stakeholders' integration via supply network management, as well as enhancing collaborative partnerships, rather than adversarial relationships, through the means of a supply network risk-sharing (O'Reilly 2010).

1.3.3   Risk Management

With reference to risk-sharing, changes in the traditional risk management approach are strictly interrelated with changes in legal frameworks. Accordingly, Chong et al. (2017) identify three main legal aspects affected by digitalization: (1) Contract structure and policy; (2) Contractual relationships and obligations; and (3) BIM model and security.

In relation to the first aspect, it is increasingly acknowledged that traditional contracts and legal frameworks are conducive to adversarial behaviors between project partiers. Concerning this matter, Chong and Phuah (2013) argue that traditional legal frameworks are designed to govern a sequential, silo-based process. Furthermore, Chong et al. (2017) highlight how BIM contract protocols are usually just added as appendixes to the traditional contract. In light of this, Kuiper and Holzer (2013) underline the need for rethinking the contractual context, in terms of both structure and policies. Standard contracts should be developed in accordance also to the emergent procurement system, such as Integrated Project Delivery (IPD).

Although the most common delivery method, i.e. the Design–Bid–Build approach[4] (DBB), is still widely adopted, other innovative models, such as IPD[5]—but also Public-Private Partnerships (P3s)—are starting to receive consideration. The relational approach which underlies these new procurement methods—as opposed to the traditional, adversarial one (e.g. DBB, etc.)—is based upon a risk-and-reward sharing principle. As the PPP Canada website states "Risks arise in all projects, regardless of the procurement approach. In a P3, project risks are transferred to the party best able to manage them. By making the private sector responsible for managing more risk, governments reduce their own financial burden. The private sector bids a fixed price for the bundled contract and must pay out of pocket should any unforeseen expenses arise (e.g. cost escalation, construction defects, unexpected maintenance requirements, etc.)."

Regarding the second aspect proposed by Chong et al. (2017), namely 'contractual relationships and obligations', Sebastian (2011) highlights how the industry digitalization is changing relationships between project parties. Architects, consulting engineers, contractors, sub-contractors, etc. are more and more required to partner and work collaboratively (Poirier et al. 2017). As Biancardo et al. state (2015), "[…] They are all shifting away from the traditional paradigm, one that places users, planners, designers, and contractors in different silos during professional practice".

Given the relational work context brought on by digital technologies and processes, roles, responsibilities, and relationships should be clearly defined to avoid legal issues. Some documents, as the BIM Execution Plan, may help to manage this issue. However, these documents are usually not included in the contract (Hardin and McCool 2015). Regarding this matter, McAdam (2010) points out the importance of defining contractual relationships among project parties. A clear definition of contractual relationships may also foster other positive trends, such as: performance-based honorarium system (Sebastian 2011), design liability (Hayne et al. 2014; Sieminski

---

[4]This is a contract framework wherein the owner has a contract with the architect or engineer to design the work, and a separate contract with a contractor to construct the design. Additionally, the general use of "contract" may include examples or discussion of private and/or government contracts, and/or any number contractual arrangements, including, for example, lump sum payment, guaranteed maximum price, and so on. Likewise, "design professional" may be used to refer to architects and/or all form of consulting engineers (mechanical, electrical, civil, etc.) with "contractor" similarly referring to a general contractor or construction manager of various type, unless specifically noted otherwise.

[5]The AIA California Council (Sieminski 2007) *Guide* proposes the following definition of IPD: "Integrated Project Delivery (IPD) is a project delivery approach that integrates people, systems, business structures and practices into a process that collaboratively harnesses the talents and insights of all participants to optimize project results, increase value to the owner, reduce waste, and maximize efficiency through all phases of design, fabrication, and construction. IPD principles can be applied to a variety of contractual arrangements and IPD teams can include members well beyond the basic triad of owner, architect, and contractor. In all cases, integrated projects are uniquely distinguished by highly effective collaboration among the owner, the prime designer, and the prime constructor, commencing at early design and continuing through to project handover".

2007), and intellectual property rights and security issues[6] (Mahamadu et al. 2013; Singh et al. 2011).

A lacking definition of stakeholders' roles, responsibilities, and relationships is not only risky in terms of legal issues (Chong et al. 2017), but it can also hinder the optimal implementation of a performance-based design process (Sebastian 2011). Furthermore, Sebastian (2011) argues that a performance-based design process can be enhanced by a performance-based honorarium system. However, the traditional payment schemes are usually calculated as a percentage of project costs: the more expensive the building is, the higher the honorarium will be. This way, when issues arise, adversarial behaviors are more likely to be expected. On the contrary, integrated procurement systems, such as IPD, are more conducive to create a performance-based and non-adversarial design process. Since project parties receive an honorarium or payment according to the building quality over a certain period, they are more incentivized to achieve an optimal design quality. A performance-based honorarium system can stimulate stakeholders to continuously assess the life-cycle consequences of their project solutions. To conclude, as Sebastian (2011) argues, *"Using BIM substantially alters the relationships between parties and blends their roles and responsibilities [...] As we move forward with BIM projects, risks will need to be allocated rationally, based on the benefits a party will be receiving from BIM, the ability of the party to control the risks, and the ability to absorb the risks through insurance"*.

References

Al Ahbabi M, Alshawi M (2015) BIM for client organisations: a continuous improvement approach. Constr Innov 15(4):402–408

Allen T (1977) Managing the flow of technology. MIT Press, Cambridge, MA

Biancardo SA, Osmanbhoy N, Ottesen JL, Migliaccio GC, Clevenger C (2015) Closing the contractual circle: investigating emergent subcontracting approaches. In: Proceedings of the 5th international/11th construction specialty conference. Vancouver, British Columbia

Burt R (1992) Structural holes. Harvard University Press, Cambridge, MA

Chong HY, Wang J, Shou W, Wang X, Guo J (2014) Improving quality and performance of facility management using building information modelling. In: Luo Y (ed) Cooperative design, visualization, and engineering. CDVE 2014. Lecture notes in computer science, vol 8683. Springer

Chong H, Fan S, Sutrisna M, Hsieh S, Tsai C (2017) Preliminary contractual framework for BIM-enabled projects. J Constr Eng Manage 143(7)

Chong HY, Phuah TH (2013) Incorporation of database approach into contractual issues: methodology and practical guide for organizations. Autom Constr 31:149–157

---

[6]Information included in BIMs must not only be associated to the responsible project parties, but also protected when shared with others. The intellectual property rights need to be defined at the early stage of project development via BIM contract protocols (e.g. ConsensusDOCS 301 BIM Addendum, AIA Document E202, etc.) (Mahamadu et al. 2013). As Chong et al. (2017) pointed out, information in BIMs are digitized; in practical terms, this means that they can be easily extracted and reused. Therefore, new questions arise, namely: how the business knowledge can be protected? Contractual relationships may help to manage these kinds of issues too.

Czmocha I, Pekala A (2014) Traditional design versus BIM based design. Procedia Eng 91:210–215

Deutsch R (2015) Data-driven design and construction: 25 strategies for capturing, analyzing and applying building data, 1st edn. Wiley, Hoboken, NJ

Duhon B (1998, Sep) It's all in our heads, Inform, vol 12, No 8, pp 8–13

Eastman C (2011) BIM handbook: a guide to building information modeling for owners, managers, designers, engineers and contractors, 2nd edn. Wiley, Hoboken, NJ

Eizenman E, Khan N, Schrey C (2017) How digital is changing leadership roles and responsibilities. Digital McKinsey. McKinsey & Company, New York

Gerbert P, Castagnino S, Rothballer C, Renz A, Filitz R (2016) Digital in engineering and construction. the transformative power of building information modelling. The Boston Consulting Group Inc., Boston, MA

Hanifan G, Sharma A, Newberry C (2014) The digital supply network: a new paradigm for supply chain management. Accenture Strategy

Haponava T, Al-Jibouri S (2010) Establishing influence of design process performance on end-project goals in construction using process-based model. Benchmarking Int J 17(5):657–676

Hardin B, McCool D (2015) BIM and construction management: Proven tools, methods, and workflows, 2nd edn. Wiley, Hoboken, NJ

Hayne G, Hare B, Kumar B (2014) The development of a framework for a design for safety BIM tool. In: Proceedings of the 2014 international conference on computing in civil and building engineering

Hegazy FM, Ghorab KE (2014) The influence of knowledge management on organizational business processes' and employees' benefits. Int J Bus Soc Sci 5(1):148–172

Kuiper I, Holzer D (2013) Rethinking the contractual context for building information modelling (BIM) in the Australian built environment industry. Aust J Constr Econ Build 13(4):1–17

Lindhard S, Larsen JK (2016) Identifying the key process factors affecting project performance. Eng Constr Arch Manage 23(5):657–673

Mahamadu AM, Mahdjoubi L, Booth C (2013) Challenges to BIM-cloud integration: implication of security issues on secure collaboration. In: 2013 IEEE international conference on cloud computing technology and science. IEEE, New York, pp 209–214

McAdam B (2010) Building information modelling: the UK legal context. Int J Law Built Env 2(3):246–259

McGraw Hill (2014) Smart market report, the business value of BIM for owners. McGraw Hill Construction, Bedford, MA

Mussomeli A, Gish D, Laaper S (2016) The rise of the digital supply network. Industry 4.0 enables the digital transformation of supply chains. Deloitte University Press

O'Reilly D (2010) Team approach, risk sharing necessary for BIM. J Commer

Papadonikolaki E, Vrijhoef R, Wamelink H (2016) The interdependences of BIM and supply chain partnering: empirical explorations. Arch Eng Des Manage 12(6):476–494

Papadonikolaki E, Wamelink H (2017) Inter- and intra-organizational conditions for supply chain integration with BIM. Build Res Inf 45(6):649–664

Picon A (2016) From authorship to ownership. Arch Des 86(5):36–41

Poirier EA, Forgues D, Staub-French S (2017) Understanding the impact of BIM on collaboration: a Canadian case study. Build Res Inf 45(6):681–695

Rogers EM (1995) Diffusion of innovations. Free Press, New York

Schrage M (2016) Instead of optimizing processes, reimagine them as platforms. Harward Business Review, December 2016

Sebastian R (2011) Changing roles of the clients, architects and contractors through BIM. Eng Constr Arch Manage 18(2):176–187

Sieminski J (2007) Liability and BIM. AIA best practices. Available at: www.aia.org/aiaucmp/groups/ek_members/documents/pdf/aiap037060.pdf

Singh V, Gu N, Wang XY (2011) A theoretical framework of a BIM-based multidisciplinary collaboration platform. Autom Constr 20(2):134–144

Succar B (2010) The five components of BIM performance measurement. In: Proceedings of CIB world congress, Salford

Tassera A (2017) Data is changing the way we understand, design & live in cities. Available at: https://www.linkedin.com/pulse/data-changing-way-we-understand-design-live-cities-andrea-tassera

Vasudevan A, Goel G, Shukla S (2018) Building a digital supply chain the right way. APICS Magazine. Available at: http://www.apics.org/apics-for-individuals/apics-magazine-home/magazine-detail-page/2018/01/04/building-a-digital-supply-chain

Walbe Ornstein S, Saraiva Moreira N, Ono R, Limongi França AJG, Nogueira RAMF (2009) Improving the quality of school facilities through building performance assessment. J Educ Adm 47(3):350–367

Zanchetta C, Croatto G, Paparella R, Turrini U (2014) Performance based building design to ensure building quality: from standardization to LEAN construction. Techne J Technol Arch Environ 8:62–69

Zhongbao L, Xiangfeng L (2013) Comparative analysis on the building design process between traditional technique and the one based on BIM technology. J Appl Sci 13:2363–2365

Chapter 2
Digital Transformation Strategies of Multidisciplinary Design Firms: Key-takeaways from Experts' Interviews

Abstract The aim of this chapter is to present the digital transformation strategies adopted by a sample of multidisciplinary design firms. For doing this, first, the author selects three Italian multidisciplinary design firms of different sizes (small, medium, and big) and at diverse digital maturity levels. Second, semi-structured interviews with the BIM manager/coordinator of each firm are conducted to collect data about the digital transformation strategy adopted by the firm. Each of the three firms is investigated according to the same pre-defined theoretical framework. For doing this, the author proposes four interpretation lenses to frame the changes which multidisciplinary design firms implement in order to manage digital transformation. These four lenses are People, Process, Product, and Policy (4Ps). Finally, the combination of the literature review and expert interviews leads the author to identify two paradigm shifts which synthesize the process-oriented and organizational changes associated with the digital transformation of multidisciplinary design firms: (1) From a silo-based and sequential to a collaborative and iterative digital design process; and (2) From a vertical and hierarchical to a platform-based and networked organization.

Keywords Digital transformation · Multidisciplinary design firms · Design process · Collaboration · Organizational change · Platform · Network

2.1 People, Process, Product, and Policy (4Ps): The Interpretation Lenses of Digital Transformation

In addition to the literature review about the emergent digital transformation trends (see Chap. 1), this study proposes an empirical level of investigation performed through semi-structured interviews with A/E industry experts. The purpose is to understand how digital transformation impacts processes, people, products (in terms of software and tools), and policies of multidisciplinary design firms.

To achieve this, the author of this study selects three firms as a representative sample of Italian multidisciplinary design firms in a period of digital transformation (see Table 2.1). Norsa (2017) proposes a list of Italian "BIM ready" design firms. The selection is made in relation to the presence (or not) of either BIM Managers

Table 2.1 Profile of design firms selected as case studies for the scope of this study

Firm	Type of design firm	Size of firm (by no. of employees)[a]	Approximate years of BIM usage experience	Role of interviewee within the firm
A	Multidisciplinary design firm	Medium	5 years	BIM coordinator
B	Multidisciplinary design firm	Large	8 years	BIM manager and deputy BIM manager
C	Multidisciplinary design firm	Small	4 years	BIM manager

[a]Firm size: micro <10, small <50 employees, medium <250 employees, and large $\geq$250 employees (European Commission 2005)

or BIM Coordinators within the firm's organizational structure. This list includes 36 firms spread across different Italian locations. Each of the three firms selected for this study is included in the list proposed by Norsa (2017). It can be reasonably argued that the selected sample represents in a convenient way the Italian multidisciplinary design firms in periods of digital transformation.

As shown in Table 2.1, the firms selected vary in size and years of experience in BIM. These variations enrich the data in terms of providing the opportunity to explore differences in digital transformation strategies.

A questionnaire is designed by the author and used as a general outline to conduct the semi-structured interviews with the BIM Manager/Coordinator of each of the selected firms. Before being used, the questionnaire is reviewed by both researchers and practitioners for enhancement in terms of content validity. The proposed questionnaire contains a set of pre-defined topics with a pre-determined order and it is divided into four sections: (1) People; (2) Process; (3) Product; and (4) Policy (4Ps).

1. The 'People' section aims at covering the following topics:

    - Personnel training. Specifically, questions are formulated to know if and how:
        - The training is managed internally or externally;
        - Both the operative (power workers) and the managerial (project managers/partners) personnel is involved;
        - All the business units have been trained since the early stages of the firm's digital transformation.
    - Changing and new roles and responsibilities. Specifically, questions are defined to know if and how:
        - New roles and responsibilities are formally defined;
        - Existing organizational structure is (re)-configured accordingly to new or changing roles and responsibilities.

2. The 'Process' part investigates if and how:

- Supply chain relationships are changed. Specifically, questions are posed to learn if and how:
  - Consultants and suppliers are chosen in relation to their digital skills and capabilities;
  - (Im-)maturity of the reference market in terms of digital technologies and processes influences the firm's digital transformation.

3. The 'Product'-oriented questions focus on how:

   - Setting the technological infrastructure, in terms of:
     - Hardware and software;
     - Standards and procedures developed to support the implementation of digital technologies and processes.

4. The 'Policy' section explores if and how:

   - Contractual and legal aspects are managed accordingly to the new requirements of digital technologies and processes;
   - Relationships with clients/owners are managed to provide digital deliverables.

2.1.1 Digital Transformation Strategy of Firm A, Milan, Italy

People

A middle-out[1] strategy is employed to manage the firm's digital transformation. In other words, a strong commitment by the top management is coupled with a bottom-up diffusion mechanism.

Internal and continuous training programs are set up, but not all the firm's business units are fully digitally-capable yet. The technical personnel (e.g. IT business unit) is involved in the training program too. However, some difficulties are faced with project managers' involvement. As the BIM Coordinator underlines during the interview, it is hard to make project managers committed to change. Nevertheless, given the importance of project managers' involvement, the strategy is to train personnel at the bottom of the authority chain to gradually diffuse the innovation up the organizational structure (bottom-up approach).

Roles and responsibilities of digitally-capable personnel are not formally defined yet.

---

[1] A middle-out diffusion combines both a top-down and a bottom-up approach. This pattern when coupled with top-management sponsorship can guarantee successful innovation adoption and knowledge creation enabled by the technology infrastructure. At the organizational level, team managers and business unit's heads boost innovation up and down the authority chain. A middle-out pattern can be the best way to transitioning to digital and BIM, if a digital business unity—either a digital group or manager—is appointed to wide-spread innovation within the firm. Effective diffusion of innovation must be coupled with a middle and top management support.

Process

Relationships with consultants and suppliers have not changed since the firm's digital transformation started. Additionally, consultants and suppliers are not chosen according to their digital skills and capabilities. Trusted consultants are still the same, even if they do not have any digital skills and capabilities and they are not even 'pushed' to acquire them.

Product

The technology infrastructure (in terms of software and tools) is set up by the BIM Coordinator, who is also in charge to gradually implement it in accordance with the firm's digital maturity level.

Policy

Relationships with clients have not changed since the firm's digital transformation started. Specific contractual relationships to support the use of digital technologies and processes are defined only if the client requires them. In most of the cases, the client does not even know that the firm uses digital technologies and processes to develop the project: it is an organizational choice to improve the efficiency and quality of the project.

2.1.2　Digital Transformation Strategy of Firm B, Milan, Italy

People

A top-down[2] implementation strategy is adopted by firm B, which means that the initiative came from top-management. The firm employs a multidimensional approach to the firm's digital transformation. In other words, the top-management formal commitment is coupled with people training, projects piloting, and economic investment.

Internal training programs are continuous and involve all the firm's business units. Two different training programs are set up: one to learn how to 'use' digital technologies (for project team members), another one to learn how to manage digital processes (for project managers and partners). Roles and responsibilities are formally defined within the firm's organizational structure. BIM Coordinators work side by side with project managers, while the BIM Manager holds a more managerial and strategic role. Furthermore, there is no overlap between the management of projects (led by project managers) and the management of digital transformation (led by BIM Manager/Coordinators).

---

[2]In the case of a top-down diffusion, the initiative within a firm (irrespective of its size and location within the supply chain) comes from top-management. Through this 'mandate', the innovative solution starts diffusing down the organizational structure and—if coupled with economic investment, people training and projects piloting—is fast adopted. People in power set the pace, define targets and objectives, as well as provide necessary funding for innovation adoption. Successful adoption is highly dependent on the degree, stability and wisdom of top-management support. It is acknowledged that a top-down approach within an organizational setting is the fastest way to wide-spread adoption.

Process

Relationships with suppliers and consultants have changed since the early phases of the firm's digital transformation. Consultants are chosen according to their digital skills and capabilities. If they are not digitally-capable yet, they are asked to acquire the necessary skills and capabilities in order to be part of the project. Trusted consultants are involved since the beginning of the firm's digital transformation.

Relationships with clients/owners have been reconfigured. Given the importance of clients' involvement, firm's leaders and partners always try to make the client aware of both the technology and methodology required and implemented to support digital processes.

Product

The technology infrastructure, in terms of software and hardware, is set up, as well as business standards, templates, and object libraries. The BIM Manager is in charge to gradually implement the entire infrastructure in accordance with the firm's digital maturity level.

Policy

The importance of defining contractual relationships to support the implementation of digital technologies and processes is significantly considered. The legal framework between project parties is defined *ex-ante*. Before signing the contract, the BIM Manager develops a BIM Execution Plan, which details the project development process from beginning to end.

2.1.3 Digital Transformation Strategy of Firm C, Bologna, Italy

People

A middle-out approach is employed to manage the firm's digital transformation. The process is supported by the firm's top-management and coupled with a wide diffusion at the bottom of the authority chain.

Personnel training is managed by the BIM Manager. It can be defined as 'a learning by doing' process, in other words, personnel is training while working on projects. Personnel training has started from the core business, which is the architecture business unit, and then involved all the other ones, such as the structural engineering business unit. Trusted consultants have also been gradually involved in this training process.

Project teams are strategically created by combining together people with and without digital skills and capabilities. Furthermore, different training programs are set up to provide 'power workers' and project managers with the necessary skillset. Senior and top management is involved in setting up the training process. The motivation is the need for considering the features of the traditional way of working, and therefore the existing quality management system, to define the new one. Roles and responsibilities are not rigorously defined yet. Project managers are usually flanked by the BIM Manager throughout the entire digital project development.

Process

Trusted consultants and suppliers have been informed of the firm's digital transformation since the early phases. They have also been suggested to take part in this transformation. The first digital pilot project was carried out with their collaboration. Afterward, firm C has started to open new channels by collaborating also with other partners and consultants, who have been selected according to their digital skills and capabilities.

Clients/owners are involved in the firm's digital transformation since the firm always tries to make clients understand the new business opportunities that digital technologies and processes can bring to them. However, they have never delivered digital models to any client so far. When it happened that the firm was required by the client to develop a project within a digital environment, it was always an international client, sometimes with a business branch in Italy. As the firm's BIM manager stated during the interview (2016) "Due to the importance of clients' involvement in a digital design process, designers have the responsibility to educate them about playing an active and digitally-conscious role throughout the entire process".

Product

The technology infrastructure is set up, as well as business standards and graphical templates. The BIM Manager is in charge to gradually implement the entire infrastructure in accordance with the firm's digital maturity level. Furthermore, a digital team, which works in collaboration with academia, is responsible to explore new software and tools, as well as to publish articles and white papers about emergent trends and innovative scenarios.

Policy

No information has been provided about this category.

2.2 Process-Oriented and Organizational Paradigm Shifts of Digital Transformation

The combination of literature review (see Chap. 1) and experts' interviews has led the author of this study to identify two paradigm shifts of digital transformation from the perspective of process and organization: (1) From a silo-based and sequential to a collaborative and iterative digital design process; and (2) From a vertical and hierarchical to a platform-based and networked organization.

2.2.1 From a Silo-Based and Sequential to a Collaborative and Iterative Digital Design Process

Given the challenges which have been traditionally associated with the design process management, a need for change is increasingly emerging. In the traditional design process, in fact, tasks are managed as silos, rather than as systems, and decisions are exchanged between project stakeholders from different disciplines only after having been taken. This indicates a lack of cross-functional collaboration between project parties. Furthermore, project relationships are often defined by adversarial transactions of work, rather than by collaborative interactions on the project. Project parties, in fact, often wait for each other's completion of deliverables. This way of working shows a lack of iteration throughout the project development phases (Zhongbao and Xiangfeng 2013).

This lack of collaboration and iteration determines a low performance in terms of process efficiency and product quality (Lindhard and Larsen 2016). Furthermore, the silo-based and sequential nature of the traditional design process contributes to the creation of an adversarial environment. Jaffar et al. (2011) identify five factors of conflict, which characterizes the AEC industry: (1) Increasing project cost; (2) Project delays; (3) Reduced productivity; (4) Loss of profit; and (5) Damage in business relationships. They also classify recurring problems of the traditional design process into three categories: behavioral, contractual, and technical problems. Cakmak and Cakmak (2014) contribute to this topic by classifying the causes of disputes as (A) owner related, (B) contractor related, (C) design related, (D) contract related, (E) human-behavior related, (F) project related, and (G) external factors.

Additionally, if we consider that the design process involves many stakeholders from different disciplines, it is a matter of fact that design firms deal with a large amount of data and information provided by various sources and in many forms. Communication and information exchange are therefore often characterized by a high level of complexity. The lack of collaboration between project parties and of iteration throughout process tasks determines problems, such as information asymmetry and data fragmentation.

Furthermore, if we consider also the lack of interoperability between tools, it can be understood why reciprocal interdependencies between project parties play such an important role in terms of process efficiency and product quality. Specifically, inadequate interoperability represents a cost burden for all the stakeholders involved in the construction process. It has been estimated that the cost of inadequate interoperability in the U.S. market, for example, is equal to $15.8 billion per year (see Table 2.2). Architects and engineers bore approximately $1.7 billion (Gallaher et al. 2004).

The importance of a shift from a silo-based and sequential to a collaborative and iterative digital design process (Poirier et al. 2015) can be further demonstrated by mentioning two emergent digital transformation trends (see Chap. 1). Namely:

- **Design with data for empowering 3D with D³ (Data-Drive Decisions)**. This shift to data-driven decision making requires iteration throughout project phases and

Table 2.2 Costs of inadequate interoperability classified by stakeholder group and across life-cycle phase (Gallaher et al. 2004)

Stakeholder group	Planning, design, and engineering, phase	Construction phase	Operations and maintenance phase	Total
Architects and engineers	1,007.2	147.0	15.7	1,169.8
General contractors	485.9	1,265.3	50.4	1,801.6
Specialty fabricators and suppliers	442.4	1,762.2	–	2,204.6
Owners and operators	722.8	898.0	9,027.2	10,648.0
Total	2,658.3	4,072.4	9,093.3	15,824.0

collaboration between project parties. This way, different project alternatives can be collaboratively evaluated to achieve the optimal process performance and product quality according to a set of relevant project parameters (Zanchetta et al. 2014);

– *Network approach: an ongoing revolution in project stakeholder management.* Collaboration among project parties is to be considered as an enabling factor for this scenario. The goal is to improve process performance and product quality by shifting from supply chains to digital supply networks and from adversarial relationships to collaborative partnerships between project parties.

Additionally, the changes as reported by the experts' interviews contribute to outline collaboration and iteration as emergent paradigms of digital transformation from the perspective of the process. For example, interviews with industry experts underline that temporary and project-based relationships between parties are being transformed into long-term and firm-based partnerships with the extent to involve all the project parties in the definition and management of digital design processes. Regarding this matter, it can be reminded what the BIM manager of firm C stated (2016): "Due to the importance of clients' involvement in a digital design process, designers have the responsibility to educate them about playing an active and digitally-conscious role throughout the entire process".

2.2.2 From a Vertical and Hierarchical to a Platform-Based and Networked Organization

As the design process goes through a paradigm shift, so organizational structures must do (Gerbert et al. 2016; Picon 2016). The motivation is that vertical and hierarchical organizations do not facilitate efficient and effective knowledge management, in terms of creation, distribution, and communication (Allen 1977, Burt 1992, Rogers 1995).

This shift has been identified by focusing on the other digital transformation trend investigated in Chap. 1. Namely:

– *Platform-based and networked organizational structures for leveraging the power of ecosystems*. In this regard, it should be noted that platform-based organizational structures can leverage network effects. Therefore, the idea is to shift from traditional silos to innovative ecosystems by enhancing networked business communities of practice.

Furthermore, the changes reported in the expert interviews contribute to identify dynamicity and network as emergent paradigms of the organizational structures in the context of the digitalization. For example, the expert interviews reported in the previous sections have shown how organizational silos are being transformed into ecosystems by enhancing networked business communities aimed at improving cross-functional knowledge management.

2.2.3 Why Study the Digital Transformation of Multidisciplinary Design Firms from the Perspectives of Process and Organization?

In a recent study conducted by the Association of German Chambers of Commerce and Industry (DIHK), 93% of the interviewed firms argue that digitalization has impacts on their business processes and business models (Gerbert et al. 2016). Another recent study, conducted by the MIT Sloan Management Review in collaboration with Deloitte (Kane et al. 2016), argues that firms should reconfigure both business processes and business models to lead digital transformation successfully. In this report, entitled "Aligning the Firm for its Digital Focus", more than 3,700 survey participants[3] reply as follows:

– Nearly 90% of this sample agree that digitalization affects processes and organizational structures;
– However, just 44% of the survey participants think that their firms are prepared to manage process innovation and organizational structure reconfiguration (Kane et al. 2016).

These data show that firms are increasingly aware of the transformational power of digitalization, but they do not feel prepared to manage the associated process-oriented and organizational changes. Implementing a firm's digital transformation, in fact, is a complex process due to the multidimensional nature of the changes

---

[3]The survey, conducted in 2015, covered a sample of 131 countries and 27 industries. In addition to survey results, business executives from many industries, as well as technology vendors, were interviewed to understand the practical issues facing firms today.

required (Azhar et al. 2012; Yan and Demian 2008). Additionally, the specificity of the different contexts of change (e.g. reference market, etc.) should be considered. There can be risks, in fact, associated with the implementation of digital technologies and processes, which can differ a lot in diverse contexts of change. Chien et al. (2014) classify these risks into six categories:

1. Technical risk;
2. Management risk;
3. Financial risk;
4. Legal risk;
5. Environmental risk;
6. Political risk.

Given these considerations, multidisciplinary design firms should develop digital transformation strategies, which take into account both the process-oriented and organizational change, as well as the specificity of their context of change. Regarding this matter, Kane et al. (2016) describe digitally-mature firms as follows: "An expanded appetite for risk, rapid experimentation, heavy investment in talent, and recruiting and developing leaders who excel at soft skills". Specifically, the authors argue that firms should boost the so-called 'digital congruence'. It means that process tasks, people minds, and organizational structures should be aligned in order to implement a successful digital transformation. "When culture, people, structure, and tasks are firing in sync, businesses can move forward successfully and confidently" (Kane et al. 2016).

References

Allen T (1977) Managing the flow of technology. MIT Press, Cambridge, MA

Azhar S, Khalfan M, Maqsood T (2012) Building information modeling (BIM): now and beyond. Australas J Constr Econ Build 12(4):15–28

Burt R (1992) Structural holes. Harvard University Press, Cambridge, MA

Cakmak E, Cakmak PI (2014) An analysis of causes of disputes in the construction industry using analytical network process. Procedia Soc Behav Sci 109:183–187

Chien KF, Wu ZH, Huang SC (2014) Identifying and assessing critical risk factors for BIM projects: empirical study. Autom Constr 45:1–15

European Commission (2005) The new SME definition. European Commission. Available at http://ec.europa.eu/enterprise/policies/sme/facts-figures-analysis/sme-definition/

Gallaher MP, O'Connor AC, Dettbarn JL Jr., Gilday LT (2004) Cost analysis of inadequate interoperability in the U.S. capital facilities industry. NIST GCR 04-867, Gaithersburg

Gerbert P, Castagnino S, Rothballer C, Renz A, Filitz R (2016) Digital in engineering and construction. The transformative power of building information modelling. The Boston Consulting Group Inc., Boston, MA

Jaffar N, Abdul Tharim AH, Shuib MN (2011) Factors of conflict in construction industry: a literature review. Procedia Eng 20:193–202

Kane GC, Palmer D, Phillips AN, Kiron D, Buckley N (2016) Aligning the organization for its digital future. MIT Sloan Management Review. Available at: https://sloanreview.mit.edu/projects/aligning-for-digital-future/

Lindhard S, Larsen JK (2016) Identifying the key process factors affecting project performance. Eng Constr Arch Manage 23(5):657–673

Norsa A (ed) (2017) Report 2016 on the Italian architecture and engineering industry. Guamari

Picon A (2016) From authorship to ownership. Arch Des 86(5):36–41

Poirier EA, Staub-French S, Forgues D (2015) Investigating model evolution in a collaborative BIM environment. In: Proceedings of the 5th international/11th construction specialty conference, Vancouver, British Columbia

Rogers EM (1995) Diffusion of innovations. Free Press, New York

Yan H, Demian P (2008) Benefits and barriers of building information modelling. In: Proceedings of the 12th international conference on computing in civil and building engineering (ICCCBE XII) and 2008 international conference on information technology in construction (INCITE 2008), Beijing, China

Zanchetta C, Croatto G, Paparella R, Turrini U (2014) Performance based building design to ensure building quality: from standardization to LEAN construction. Techne J Technol Arch Env 8:62–69

Zhongbao L, Xiangfeng L (2013) Comparative analysis on the building design process between traditional technique and the one based on BIM technology. J Appl Sci 13:2363–2365

Chapter 3
Paradigms, Perspectives, and Context of Change for the Digital Transformation of Multidisciplinary Design Firms

Abstract The aim of this chapter is to propose a framework to identify, structure, and categorize the changes brought on by digital transformation within multidisciplinary design firms. This framework consists of three conceptual categories: paradigms of change, perspectives of change, and context of change. In accordance with the paradigm shifts outlined in Chap. 2, the paradigms of change proposed are collaboration, iteration, platform, and network. These paradigms of change are coupled and cross-combined with the perspectives of change: process and organization. Additionally, this study underlines the importance of considering the specificity of the context of change, which consists of the combination of three elements: reference market/industry, local and national institutions and related standards available, and the specific firm's characteristics.

Keywords Digital transformation · Multidisciplinary design firms · Design process · Collaboration · Organizational change · Platform · Network

3.1 Paradigms of Change: Collaboration, Iteration, Platform, and Network

As described in Chap. 2, the paradigm shifts that outline the process-oriented and organizational changes in multidisciplinary design firms in periods of digital transformation are (1) From a silo-based and sequential to a collaborative and iterative digital design process and (2) From a vertical and hierarchical to a platform-based and networked organization. The paradigms of change highlighted by these shifts—collaboration, iteration, platform, and network—are coupled and cross-combined with the proposed perspectives of change, namely process and organization, in this way:

– Collaboration and iteration from the design process perspective;
– Platform and network from the organizational perspective.

Collaboration, iteration, platform, and network are therefore proposed by this study as the paradigms of change through which to manage design process innovation and organizational structure reconfiguration in multidisciplinary design firms in periods of digital transformation.

3.1.1 Collaboration and Iteration

The literature review proposed in Chap. 1 and the expert interviews reported in Chap. 2 show the increasing call to boost value creation in the industry. The mainstream silo-based and sequential approach to design process management (Cakmak and Cakmak 2014; Jaffar et al. 2011) is increasingly challenged by the need for more collaboration between project parties (Sebastian 2011) and iteration across project phases (Elmualim and Gilder 2014).

With reference to the concept of collaboration, Hughes et al. (2012) propose the following definition: "Collaboration is a non-adversarial team-based environment, where through the early involvement of key members and the use of the correct contract, everyone understands and respects the input of others and their role and responsibilities. The relationships are managed with the help of regular meetings, early warning systems, open dialogue and risk sharing to produce an atmosphere of mutual trust where information is shared, problems can be solved together, and everyone contributes towards a common aim motivated by a fair method of pain-share gain to produce a win-win outcome." This definition of collaboration underlines the importance of people-oriented factors, such as the setting of a 'non-adversarial team-based environment' and process-related aspects, e.g. the 'the help of regular meetings, early warning systems' to facilitate collaboration between project stakeholders.

Most failures are experienced due to digital transformation strategies that mainly focus on the implementation of new software and tools and overlook the changes in the process and organizational setting (Erdogan et al. 2014). "The success of a collaboration environment not only depends on the collaboration technology introduced to the organization but is also highly influenced by how this technology is introduced. Unless supported by the relevant people, process, and change management strategies, the stand-alone implementation of collaborative information technology (IT) will not enhance collaborative working." (Erdogan et al. 2014). Accordingly, Dossick and Neff (2011) point out the 'mistake' of managing a firm's digital transformation by primarily focusing on technical issues associated with interoperability and data exchange. The authors carried out many practice observations and interviews, which led to demonstrate that setting up an innovative technological infrastructure is not sufficient to manage collaboration challenges.

Regarding the concept of iteration, Saki and Stravoravdis (2015) argue that "Iteration is defined as a process of repeating a set of steps until a desirable result is achieved and any design process is by nature iterative. In order to improve the specification of a product, it is crucial to generate as many design concepts as possible at the earliest stages and to evaluate them and prepare feedback." However, professionals typically work in sequential work transactions rather than collaborative interactions to explore and evaluate many design alternatives.

Although there is much research on digital transformation, innovation is still suboptimal. This study argues that one motivation is in the imbalanced focus put on the 4Ps lenses of change: People, Process, Product, and Policy. Today, a 'close approach' to digital transformation that mainly depends on a certain software application is still widely implemented in the industry. In this way, some of the lenses of change associated with the digital transformation are overlooked: much time is spent to define the new technological infrastructure, and less on the reengineering of processes and the reconfiguration of organizational structures.

3.1.2 Platform and Network

In addition to collaboration and iteration, the analysis of both the literature review and experts' interviews has led to identify the other two paradigms of change: platform and network. The traditional vertical and hierarchical organizations (Cakmak and Cakmak 2014; Jaffar et al. 2011) are increasingly challenged by the emergent platform-based (Picon 2016) and networked (Schrage 2016) new forms of organization. Regarding these new forms of organization, Picon (2016) argues that "The generalization of the use of the computer and the triumph of the Internet have fostered innovative design practices, the networked structure of which breaks from traditional top-down organizations. A new economy of the architectural field is emerging in which information circulates in a manner that evokes blood flow." This statement underlines that shifting toward these new forms of platform-based and networked organizations leads to improve information management too.

To conclude, this study therefore argues that, as design firms transition to collaborative and iterative digital design processes, their organizational structures must be reconfigured accordingly in order not only to align with the new and changing roles and responsibilities associated with digital work environments but also to support an efficient and effective knowledge management (Allen 1977; Burt 1992; Rogers 1995).

3.2 Perspectives of Change: Process and Organization

The four paradigms of change outlined in the previous section are herewith investigated from the combined perspectives of process and organization. This study argues that these two perspectives need to be considered as interacting elements, in other words, as the combined perspectives of change associated with digital transformation. Many research shows, that collaborative and iterative digital design processes are more easily implemented within platform-based and networked organizations, which foster collaborative relationships and interactions between people (Cross et al. 2013). As Cross et al. argue (2013) "Business performance is not just a function of capable people and superior practices and processes, but also the product of relationships and connections." (Cross et al. 2013).

Accordingly, many research argues that the process of innovation adoption and diffusion, as the digital transformation of multidisciplinary design firms, requires a multidimensional approach (Cooper 1998; Rogers 1995; Tornatzky and Fleisher

1990). In accordance with this view, numerous studies suggest that digital transformation has an impact on many aspects, such as technology, process, and personnel, and that these aspects should be all and equally considered. A study from Gu and London (2010) underline that there are numerous factors affecting digital transformation and group them into technical and non-technical issues. Furthermore, the authors develop a multidimensional "Collaborative BIM Decision Framework" to facilitate BIM implementation through informed selection of *tools*, based upon *people* readiness and capabilities, and *process* dependencies. Accordingly, Staub-French et al. (BIM Topics Lab) propose a conceptual framework to investigate BIM from the perspectives of Technology, Organization, and Process (the three primary dimensions) in Context (industry, region, culture, etc.) and across stages (adoption always occurs in stages).

Focusing on BIM as an example of digital technology and process that increasingly diffuses across the industry, literature reports two different definitions: BIM as a 'process' and BIM as a 'tool' (Deutsch 2011). Jernigan (2007) defines this dyadic nature of the understanding of BIM as "BIG BIM. little bim":

- The 'BIG BIM' approach defines BIM as a process and the acronym stands for Building Information Modeling/Management. This definition focuses on the 'process' that stands behind the development of a BIM model, and less on the 'tool' to create it;
- The 'little bim' perspective refers to BIM as a 'tool'. In this case, the acronym stands for Building Information Model, and it therefore describes BIM as a 'tool' to develop data-rich parametric models.

Considering these two different perspectives about the definition of BIM, Jernigan (2007) couples them together and defines BIM as: "A set of interacting policies, processes, and technologies generating a methodology to manage the essential building design and project data in digital format throughout the building's life-cycle." Many other researchers suggest that BIM has a multidimensional nature. Succar (2013), for example, proposes a multidimensional definition of BIM: "A set of technologies, processes, and policies enabling multiple stakeholders to collaboratively design, construct, and operate a facility." In addition to the 'process' and 'technology' perspectives, many researchers emphasize the human dimension of BIM (Reddy 2011; Succar and Kassem 2015): "BIM is about people and process as much as it is about technology" (Specialist Engineering Contractors Group 2013). Authors' belief is that a successful digital transformation requires a balance between technology, organization, and process. Furthermore, they underline the importance of cross-combining these dimensions of change with a full understanding of the context of change, as well as the stage of implementation.

All these studies argue that there are various dimensions to consider in order to implement a firm's digital transformation. Given this point of departure, we argue that a firm's digital transformation should be implemented by managing the 4Ps lenses of change: Process, People, Product, and Policy. This translates into organizational change management that does not only focus on the new technological infrastructure—in terms of software and tools—but also on redesigning processes according to the new paradigms of collaboration and iteration, as well as reconfiguring the existing organizational structure in order to shift toward a platform-based and networked organization.

3.3 Context of Change: The Interaction of Industry, Institutions, and Firm's Specificity

In addition to the understanding of the paradigms and perspectives of change, this study argues that a critical analysis of the specificity of the context of change is a fundamental step to manage a firm's digital transformation. There are many risks, in fact, associated with the potential immaturity or lack of support of the context of change. There is consequently the need for multidisciplinary design firms to analyze the specificity of their firm, reference industry/market, and institutions in order to implement risk-response organizational change management. In light of this, multidisciplinary design firms must calibrate their organizational change management according to the following factors:

- Peculiarity of the firm's processes and organizational structures;
- Readiness level of firm's personnel in terms of digital capability and expertise;
- Availability of economic resources to build up the new technological infrastructure, in terms of software and hardware;
- Digital capability of the supply chain;
- Level of maturity of the reference market in terms of digital capability and expertise;
- Availability of national standards and/or guidelines, as well as integrated delivery methods and legal frameworks.

To conclude, this study argues that the reconfiguration of organizational structures, the shift to a collaborative and iterative digital design process, as well as the (im-)maturity evaluation of the specific context of change are strategic steps to implement a firm's digital transformation.

3.4 Existing Frameworks and Methods for Change Management Associated with Digital Transformation

An increasing number of design firms aim at implementing a digital transformation and many studies have been developed to understand the associated challenges and opportunities. Diverse conceptual frameworks and methods for digital transformation have been developed so far to identify implementation factors and areas to be considered for managing successfully a firm's digital transformation, as well as to identify capability and maturity levels.

The most prolific researcher in this area is Succar (2010, 2013), Succar et al. (2012), and Succar and Kassem (2015). The approach proposed by Succar links the different digital and BIM maturity levels to be achieved with the necessary competence and skills required. As part of his approach, Succar, together with Kassem (2015), proposes a qualitative "Macro Maturity Components" model that identifies eight complementary factors for evaluating digital and BIM maturity:

1. Objectives, stages, and milestones;
2. Champions and drivers;
3. Regulatory framework;
4. Noteworthy publications;
5. Learning and education;
6. Measurements and benchmarks;
7. Standardized parts and deliverables;
8. Technology infrastructure.

Another noteworthy conceptual framework for BIM implementation is developed by Jung and Joo (2011). The proposed framework is based on three dimensions and six categories. Specifically, the three dimensions are BIM technology, BIM perspective, and construction business functions. The 'BIM technology' dimension is divided further into four categories: property, relation, standards, and utilization. Sebastian and Van Berlo (2010) propose the "BIM Quick Scan" tool based on a multidimensional approach to digital transformation. The "BIM Quick Scan" tool aims at 'scanning' an organization over four parameters that represent both 'hard' and 'soft' aspects of BIM, namely:

1. Organization and management;
2. Mentality and culture;
3. Information structure and information flow;
4. Tools and applications.

The scores for each parameter and the total score are visualized through a radar diagram. The underperformance of some parameters can therefore be easily identified and put into perspective and in a comprehensive relationship with the other ones.

Two elements combine all these studies together: one element is proposed as a potential and opportunity, the other one as a critical issue:

– Potential and opportunity
   Most of the frameworks and methods proposed in the literature understand and analyze digital transformation as a combination of many dimensions. Therefore, they propose to apply a multidimensional approach to digital transformation, and thus to the associated organizational change management;

– Criticalities

Most of the frameworks and methods proposed in the literature lack the integration with methodologies to organize, analyze, and visualize systematically the changes required to implement a digital transformation *in practice*. They do not focus on *how* a firm's digital transformation should be practically carried out since most of them operate only at a strategic level. Additionally, they are not fed with data about the specific context of change in order to target organizational change management accordingly.

Regarding this latter concept, many studies argue that analyzing the context before planning the change is a successful approach to organizational change management. Price and Chahal (2006) argue: "An organization's ability to effectively plan and manage change depends on how reviews and analyses take into account its existing culture. Changing and adapting the cultural paradigms usually takes considerably more time than implementing new procedures or technology."

Although many research on organizational change management has documented the importance of analyzing the context in which innovation adoption and diffusion takes place, this dimension often lacks in the literature about digital transformation and BIM implementation (Poirier et al. 2015).

3.5   Need for a Systematic Methodology for Change Management Associated with Digital Transformation

Although many research efforts have been devoted to the topic of digital transformation, however, there remain uncertainties in how to implement the associated changes *in practice*. Furthermore, few studies have investigated the specific topic of *how* to manage process-oriented and organizational changes in the context of the digitalization of the A/E industry. Additionally, there are few methodologies to investigate systematically the process-oriented and organizational changes associated with the digital transformation within the specific field of multidisciplinary design firms.

Starting from this research gap, this study aims at answering the need for a systematic methodology that can support multidisciplinary design firms in leading successfully a digital transformation being aware of *what* process-oriented and organizational changes need to be implemented and *how* to implement them in practice.

References

Abdirad H (2017) Metric-based BIM implementation assessment: a review of research and practice. Arch Eng Des Manage 13(1):52–78

Abdirad H, Pishdad-Bozorgi P (2014) Trends of assessing BIM implementation in construction research. Comput Civil Build Eng 496–503

Allen T (1977) Managing the flow of technology. MIT Press, Cambridge, MA

Alshawi M, Ingirige B (2003) Web-enabled project management: an emerging paradigm in construction. Autom Constr 12:349–364

Ashcraft H (2008) Building information modeling: a framework for collaboration. Constr Lawyer 28(3)

Burt R (1992) Structural holes. Harvard University Press, Cambridge, MA

Cakmak E, Cakmak PI (2014) An analysis of causes of disputes in the construction industry using analytical network process. Procedia Soc Behav Sci 109:183–187

Cooper JR (1998) A multidimensional approach to the adoption of innovation. Manage Decis 36(8):493–502

Cross R, Kaše R, Kilduff M, King Z, Minbaeva D (2013) Bridging the gap between research and practice in organizational network analysis: a conversation between Rob Cross and Martin Kilduff. Human Res Manage 52(4):627–644

Deutsch R (2011) BIM and integrated design: strategies for architectural practice. Wiley, Hoboken, NJ

Dossick CS, Neff G (2011) Messy talk and clean technology: communication, problem-solving and collaboration using building information modelling. Eng Project Organ J 1: 83–93

Elmualim A, Gilder J (2014) BIM: innovation in design management, influence and challenges of implementation. Arch Eng Des Manage 10(3–4):183–199

Erdogan B, Anumba C, Bouchlaghem D, Nielsen Y (2014) Collaboration environments for construction: management of organizational changes. J Manage Eng 30(3)

Gu N, London K (2010) Understanding and facilitating BIM adoption in the AEC industry. Autom Constr 19:988–999

Hughes D, Williams T, Ren Z (2012) Differing perspectives on collaboration in construction. Constr Innov 12(3):355–368

Jaffar N, Abdul Tharim AH, Shuib MN (2011) Factors of conflict in construction industry: a literature review. Procedia Eng 20:193–202

Jernigan F (2007) Big BIM, little bim: the practical approach to building information modeling; integrated practice done the right way. Site Press, Salisbury

Jung Y, Joo M (2011) Building information modelling (BIM) framework for practical implementation. Autom Constr 20(2):126–133

Love PED, Lopez R, Kim JT (2013) Design error management: interaction of people, organization and the project environment in construction. Struct Infrastruct Eng 10(6):811–820

Picon A (2016) From authorship to ownership. Arch Des 86(5):36–41

Poirier E, Staub-French S, Forgues D (2015) Embedded contexts of innovation: BIM adoption and implementation for a specialty contracting SME. Constr Innov 15(1):42–65

Price ADF, Chahal K (2006) A strategic framework for change management. Constr Manage Econ 24(3):237–251

Reddy KP (2011) BIM for building owners and developers: making a business case for using BIM on projects. Wiley, Hoboken, NJ

Rogers EM (1995) Diffusion of innovations. Free Press, New York

Saki M, Stravoravdis S (2015) Using BIM to facilitate iterative design. In: Proceedings of the conference BIM 2015, Bristol, UK

Schrage M (2016, Dec) Instead of optimizing processes, reimagine them as platforms. Harv Bus Rev

Sebastian R (2011) Changing roles of the clients, architects and contractors through BIM. Eng Constr Arch Manage 18(2):176–187

Sebastian R, Van Berlo L (2010) Tool for benchmarking BIM performance of design, engineering and construction firms in the Netherlands. Arch Eng Des Manage 6(4):254–263

Staub-French S et al. Available at http://bimtopics.civil.ubc.ca/

Specialist Engineering Contractors Group (2013) First steps to BIM competence: a guide for specialist contractors

Succar B (2010) The five components of BIM performance measurement. In: Proceedings of CIB world congress, Salford

Succar B (2013) Building information modelling: conceptual constructs and performance improvement tools. Ph.D. Thesis, University of Newcastle

Succar B, Kassem M (2015) Macro-BIM adoption: conceptual structures. Autom Constr 57:64–79

Succar B, Sher W, Williams A (2012) An integrated approach to BIM competency assessment, acquisition and application. Autom Constr 35:174–189

Tornatzky LG, Fleisher M (1990) The process of technological innovation. Lexington Books, Lexington, MA

Part II
Methodological Experimentation: Proposal of a Systematic Methodology for Change Management Associated with Digital Transformation of Multidisciplinary Design Firms

Chapter 4
A Systematic Methodology for Change Management. Proposal for a Digital Transformation Support

Abstract The aim of this chapter is to propose a systematic methodology for change management associated with the digital transformation of multidisciplinary design firms. This methodology aims at identifying, framing, and assessing the process-oriented and organizational changes occurring in multidisciplinary design firms in periods of digital transformation. Specifically, the methodology proposed includes two distinct, but complementary components: (1) A *framework* to identify, structure, and categorize the areas of transformation and the associated process-oriented and organizational changes; (2) A *set of analysis methods* for identifying, visualizing, and assessing qualitatively and quantitatively these changes. The *framework* proposed is organized as a matrix which combines the paradigms of change—collaboration, iteration, platform, and network—with the perspectives of change for digital transformation, namely process and organization. The *set of analysis methods* proposed includes both qualitative and quantitative analysis methods, in particular, process mapping, stakeholder mapping, timesheet tracking, and Social Network Analysis (SNA).

Keywords Digital transformation · Building Information Modeling (BIM) · Multidisciplinary design firms · Change management · Process mapping · Stakeholder mapping · Social Network Analysis (SNA)

4.1 Identification, Framing, and Assessment of Process-Oriented and Organizational Changes Associated with Digital Transformation: Proposal of a Systematic Methodology

The main objective of the change management methodology proposed is to identify, frame, and assess the process-oriented and organizational changes associated with the digital transformation of multidisciplinary design firms. The methodology provides multidisciplinary design firms with both a conceptual framework (see Fig. 4.1) and a set of analysis methods to identify, frame, and assess changes in practice (see Fig. 4.3). Valuable information can be extracted from performing analyses of

Fig. 4.1 Conceptual
structure of the methodology
proposed

processes and organizational structures in order to draw and calibrate digital transformation strategies accordingly. Given these considerations, the methodology proposed
aims at answering three strategic questions:

1. What process-oriented and organizational changes occur in multidisciplinary
   design firms in periods of digital transformation?
2. How to identify, frame, and assess these changes in practice?
3. How to draw and calibrate digital transformation strategies according to the
   specific context of change?

4.2 Conceptual Framework and Set of Change Analysis Methods: Components of the Methodology

The methodology proposed includes both a conceptual framework (see Fig. 4.2) and
a set of analysis methods (see Fig. 4.3) to identify, frame, and assess the process-
oriented and organizational changes associated with the digital transformation of
multidisciplinary design firms. Figure 4.1 describes the conceptual structure on the
basis of which the methodology proposed is built: the analysis and management of
change for digital transformation from the combined perspectives of process and
organization in order to move toward the new paradigms of collaboration, iteration,
platform, and network.

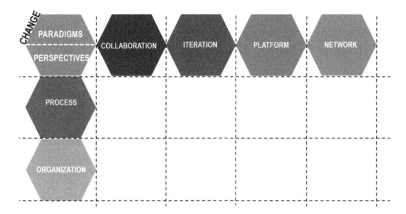

Fig. 4.2   Framework organized in paradigms and perspectives of change for digital transformation

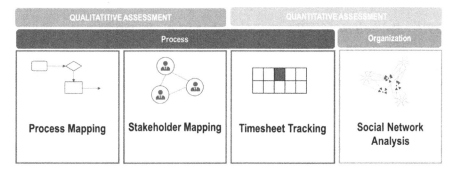

Fig. 4.3   Set of analysis methods to perform qualitative and quantitative assessments of process-oriented and organizational changes for digital transformation

4.3   Integrating Paradigms with Perspectives of Change: Conceptual Framework

The framework proposed is based on a qualitative data collection and analysis about industry trends (see Chap. 1) and best practices (see Chap. 2) for the digital transformation of multidisciplinary design firms. From a strategic point of view, the framework is intended as a conceptual structure to be used as a reference by design firms to understand areas of digital transformation. Specifically, the framework proposed aims at identifying, structuring, and categorizing changes on which to focus to implement a firm's digital transformation. Therefore, it can help organizational leaders to manage and target innovation efforts for digital transformation. From a practical point of view, it provides multidisciplinary design firms with a grid to identify the objectives of a firm's digital transformation.

The framework is built on six conceptual categories that are structured into a standardized 2 × 4 elements matrix (see Fig. 4.2). In each column, one of the following paradigms of change is entered:

– Collaboration;
– Iteration;
– Platform;
– Network.

Analogously, the two rows display the perspectives of change, namely:

– Process;
– Organization.

The elements of the so-defined matrix represent the areas of change proposed as primary to implement successfully a firm's digital transformation.

4.3.1 Paradigms of Change: Collaboration, Iteration, Platform, and Network

As previously explained, the paradigms of change for digital transformation—collaboration, iteration, platform, and network—are identified as primary by the author of this study starting from the qualitative data collection and analysis about industry trends and best practices performed via literature review (see Chap. 1) and experts' interviews (see Chap. 2). Given this theoretical background, two paradigm shifts are identified to outline the process-oriented and organizational changes occurring in multidisciplinary design firms in periods of digital transformation. Namely:

– From a silo-based and sequential to a collaborative and iterative digital design process;

"Interactive methodology design typically involves iterative process of brainstorming, prototyping, development, user testing, and evaluation. This is not a clearcut process; it often iterates through many cycles before reaching a final methodology" (Dow et al. 2005).

– From a vertical and hierarchical to a platform-based and networked organization.

"A small number of 'idea scouts' and 'idea connectors' are disproportionately influential in producing successful open innovation outcomes. Smart companies make sure they are linked" (Whelan et al. 2011).

4.3.2 Perspectives of Change: Process and Organization

The paradigm shifts proposed outline the changes associated with a firm's digital transformation from the perspectives of process and organization. These perspectives of change should not be considered as silos to be analyzed and managed separately. On the contrary, they must be considered as complementary and interacting.

Research shows that collaborative and iterative digital design processes occur more easily within platform-based and networked organizations that support efficient and effective connections and relationships for knowledge creation, distribution, and communication. As Cross et al. (2013) argue "Business performance is not just a function of capable people and superior practices and processes, but also the product of relationships and connections."

4.4 Identifying, Visualizing, and Assessing Changes: Set of Analysis Methods

Research shows that a strategic step for change management is the analysis of the context of change (Price and Chahal 2006). With reference to this topic, Price and Chahal (2006) argue: "An organization's ability to effectively plan and manage change depends on how reviews and analyses take into account its existing culture. Changing and adapting cultural paradigms usually takes considerably more time than implementing new procedures or technology."

In accordance with this view, multidisciplinary design firms should perform business assessments to bring forth a better understanding of what changes from the process and organizational perspectives occur in practice. The main objective of performing business analyses is to extract valuable information to draw and calibrate digital transformation strategies accordingly.

Starting from this point of departure, this study proposes a set of analysis methods to identify, visualize, and assess the process-oriented and organizational changes associated with the digital transformation of multidisciplinary design firms (see Fig. 4.3). To achieve this, the following methods are proposed:

- Process mapping;
- Stakeholder mapping;
- Timesheet tracking;
- Social network analysis.

The analysis methods proposed can be grouped into qualitative and quantitative. Specifically, process mapping and stakeholder mapping are proposed as methods to perform qualitative assessments, while timesheet tracking and social network analysis are proposed to carry out quantitative assessments.

4.5   As-Is, Transition, and To-Be: The Three Levels of Analysis of Change

Many studies argue that strategies for change can be diverse, but the approach based on the triangular loop 'assess - learn - implement' should always be applied. Starting from this point of departure, the set of analysis methods proposed is aimed at first assessing the traditional design process and the firm's existing organizational structure, second at learning how they change in a period of digital transformation, and lastly at envisioning the future state of both the design process and organizational structure.

Given these considerations, this study argues that the analyses to identify the process-oriented and organizational changes brought on by the digital transformation should be performed at the following three levels:

1. 'As-is' state;
2. 'Transition' state;
3. 'To-be' state.

Therefore, the three levels of analysis proposed aim at answering the following questions: (1) Where are we now? (2) How do we change? And (3) Where do we want to be?

4.6   Process Mapping and Stakeholder Mapping: Methods for Qualitative Assessment of Change

4.6.1   Process Mapping

Saluja (2009) defines process maps "[…] as tools to describe the flow of activities within the boundary of a particular topic." In general, process maps are developed by both researchers and practitioners to visualize inputs, outputs, and links of business processes (buildingSMART International Ltd. 2008). Many studies demonstrate the benefits of visualizing processes for organizations. Following are the main benefits of process mapping:

– Identifying and visualizing interdependent relationships across various process tasks, and therefore enhancing the possibility to understand them and plan accordingly;
– Understanding and communicating the divergence and disparity between the planned and executed process;
– Assessing process performance;
– Improving decision making.

Since the general purpose of process mapping is to assist in understanding and planning workflow and workforce to achieve a pre-defined objective (Wix 2007), a process map usually has a goal, specific inputs (typically from other exchange requirements and data sources), specific outputs (typically to other exchange requirements), and visualizes many activities that are performed in some order. Accordingly, Curtis et al. (1992) report that many forms of information can be extracted from a process map, namely:

- What is going to be done;
- Who is going to do it;
- When and where it will be done;
- How and why it will be done;
- Who is dependent on it being done?

As a consequence, four common perspectives of process maps are:

1. Functional: this perspective represents what process elements are performed and what flows of informational entities (e.g., data, artifacts, products, etc.) are relevant to these elements;
2. Behavioral: it mainly focuses on the time when process elements are performed and how they are performed through feedback loops, iteration, decision-making conditions, entry and exit criteria, etc.;
3. Organizational: this type of process map represent where and by whom (which agents) tasks are performed, the physical communication mechanisms used for the transfer of entities, and the physical media and locations used for storing entities;
4. Informational: this type represents the informational entities produced or manipulated by a process. These entities include data, artifacts, products (both intermediate and end products), and objects. Additionally, it includes both the structure of informational entities and the relationships among them.

In the specific context of multidisciplinary design firms, process mapping can help to understand and plan for the tasks performed within a project development process, the sequence in which these tasks are carried out, actors involved, and information exchanged between project parties (buildingSMART International Ltd. 2008). In accordance with these statements, the BIM Project Execution Planning Procedure (buildingSMART International Ltd., 2008) identifies process mapping as one of the steps for defining a BIM Project Execution Plan. The goal is to understand the implementation process for each BIM use and the implementation process of the project. Specifically, the BIM Process Mapping Procedure aims at creating a standardized procedure for planning the BIM execution on projects.

Given these considerations, process mapping is the method proposed by this study to identify, visualize, and assess the process-oriented changes brought on by a multidisciplinary design firm's digital transformation. Specifically, this study proposes to use process mapping in order to perform the following tasks:

1. Assessing the 'as-is' state, which is the traditional design process;
2. Learning how it changes because of the firm's digital transformation ('transition' state);
3. Implementing the 'to-be' state, in other words, the envisioned digital design process.

4.6.2   Stakeholder Mapping

Stakeholder mapping is typically used to analyze and visualize project parties. In general, the benefits of using stakeholder mapping can be listed as follows:

- Using opinions of project stakeholders to shape projects at an early stage. This does not only make stakeholders more likely to support the project, but their inputs can also improve project quality;
- By communicating with stakeholders early and frequently, it can be ensured that they fully understand both opportunities and challenges of the project. This means that they can provide active support and consultancy when necessary;
- Anticipating stakeholders' reactions and planning actions that will ensure their support.

For the purpose of this study, stakeholder mapping aims at understanding project parties and their relationships, as well as to analyze their digital capabilities and the potential impact on other stakeholders. The digital (im-)maturity of the stakeholders involved in a project, in fact, can be a factor of risk in periods of digital transformation. Therefore, mapping, visualizing, and analyzing project stakeholders can support the development of a digital transformation strategy, which transfers or mitigates risks upon identification.

4.7   Timesheet Tracking and Social Network Analysis: Methods for Quantitative Assessment of Change

4.7.1   Timesheet Tracking

Amongst the many uses of a project timesheet, here are some of the most important:

- Enabling tracking of the true costs of a project by accounting for the time used on it;
- Tracking the services provided by different employees;
- Comparing the hours really used on a project with the initial planning estimates;
- Automatically invoicing based on the service hours provided;
- Obtaining a list of the service hours for a given client;
- Knowing the costs needed to run the firm such as marketing costs, training costs, etc.

With reference to the specific purpose of this study, project timesheet tracking aims at helping to understand the positive and negative trends of a firm's digital transformation. Tracking the timesheet of a BIM pilot project, for example, can be useful to quantitatively assess the impact of a lack of digital skillset of project suppliers and consultants by measuring the time spent by the digitally-capable project parties to export CAD drawings from a BIM model.

4.7.2 Social Network Analysis

Social Network Analysis (SNA) aims at understanding the characteristics of organizational connections and modeling antecedents and outcomes of such relationships (Chinowsky et al. 2010, Chinowsky and Taylor 2012). In the literature, social network analysis is defined as a people-analytics method, adapted from Social Sciences, used to understand how opportunities for innovation diffuse throughout organizational connections, either formal or informal.

When considering all the dynamics of how people interact with each other within organizations, it should not be surprising that organizational 'connectors' are often not labeled as such in the formal hierarchy. Organizational structures, in fact, frequently consist of both formal hierarchy and informal networks. Specifically, the formal organizational structure includes those relationships defined by hierarchy, business processes, assigned teams, in other words, the leadership pattern put in place by management. The informal network is made up of advice and support seeking connections neither designed, nor engineered yet. Employees start creating advice and support-seeking relationships to get their job done, especially when their formal relationships do not provide them with an efficient workflow to reach the advice or support they are seeking (see Fig. 4.4).

With reference to multidisciplinary design firms in periods of digital transformation, research shows that there can be the risk that the formal hierarchy is not aligned with the informal network for digital knowledge. This lack of alignment can determine poor business performance and employees struggling for workflows leading them to the digital knowledge they are seeking. Informal networks or any underutilized potential within the formal hierarchy should be identified to restructure the organization accordingly (Cross and Parker 2004). Therefore, the analysis of both the formal relationships and informal connections is fundamental to reveal opportunities for either organizational change or integration.

Specifically, SNA aims at supporting organizational structure reconfiguration by performing the following tasks:

- Assessing the 'as-is' state of organizational structures by mapping and measuring all the formal relationships in play;
- Learning how the organizational relationships change because of the firm's digital transformation ('transition' state);
- Implementing the 'to-be' state of organizational structure on the basis of the information collected.

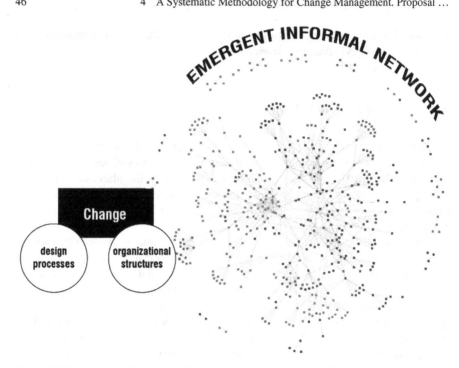

Fig. 4.4 Informal connections potentially arising as a consequence of a firm's digital transformation

Based on this, Social Network Analysis is the method proposed by this study to identify, visualize, and assess, the organizational changes brought on by the digital transformation within multidisciplinary design firms.

References

buildingSMART International Ltd. (2008) IDM: process mapping. Available at: http://iug. buildingsmart.org/idms/information-delivery-manuals/idm-for-buildingprograming/IDM_ process_map_building_programming_161111.pdf/view

Chinowsky P, Taylor J (2012) Networks in engineering: an emerging approach to project organization studies. Eng Project Organ J 2(1–2):15–26

Chinowsky P, Diekmann J, O'brien J (2010) Project organizations as social networks. J Constr Eng Manag 136(4):452–458

Cross R, Kaše R, Kilduff M, King Z, Minbaeva D (2013) Bridging the gap between research and practice in organizational network analysis: a conversation between Rob Cross and Martin Kilduff. Human Res Manag 52(4):627–644

Cross R, Parker A (2004) The hidden power of social networks: understanding how work really gets done in organizations. Harvard University Press, Boston, MA

Curtis B, Kellner MI, Over J (1992) Process modeling. Commun ACM 35(9):75–90

Dow S, Macintyre B, Lee J, Oezbek C, Bolter JD, Gandy M (2005) Wizard of Oz support throughout an iterative design process. IEEE Pervasive Comput 4(4):18–26

Price ADF, Chahal K (2006) A strategic framework for change management. Constr Manag Econ 24(3):237–251

Saluja C (2009) A process mapping procedure for planning building information modeling (BIM) execution on a building construction project. Master Thesis, Pennsylvania State University, University Park, PA

Whelan et al. (2011) Creating employee networks that deliver open innovation. MIT Sloan Manag Rev 53(1):37–40

Wix J (2007) Information delivery manual: guide to components and development methods. buildingSMART Norway. Available at: www.idm.buildingsmart.no

Part III
Applications: Test of the Methodology Proposed on Case Studies

Chapter 5
Application of the Methodology for Change Management: The Case of an Italian Multidisciplinary Design Firm

Abstract The aim of this chapter is to present the application and testing of the methodology for change management proposed in the context of an Italian multidisciplinary design firm. The objective of this firm was to move toward a collaborative and iterative digital design process. First, the conceptual framework proposed is used to identify the areas of investigation. Second, the set of change analysis methods is employed to understand the traditional design process, what process-oriented changes occur because of the firm's ongoing digital transformation, and to envision the future digital design process. To achieve this, process mapping, stakeholder mapping, and timesheet tracking are the analysis methods applied in this case study. These methods aim at identifying, visualizing, and assessing the process-oriented changes at the following three levels: (1) 'As-is' state, which refers to the design process traditionally implemented by the firm; (2) 'Transition' state, alias the process implemented to develop BIM pilot projects; and (3) 'To-be' state, which represents the envisioned digital design process. Lastly, a potential digital transformation strategy is developed in accordance with the information extracted from the performed business analyses. The strategy proposed translates into two supporting tools: (1) Strategic matrix to manage changes for digital transformation from the perspectives of process, people, product, and policy, and (2) Heat map of the digital projects developed to monitor the firm's digital transformation.

Keywords Digital transformation · Building Information Modeling (BIM) · Design process · Collaboration · Iteration · Process mapping · Stakeholder mapping

5.1 Transitioning to a Collaborative and Iterative Digital Design Process: The Firm's Objective

Firm X is one of Italy's leading architecture and engineering groups (Sole 24 Ore, 2017). The firm is organized into five brands according to a platform-based model. Each of these brands focuses on a different field: (1) Architecture and engineering, (2) Workplace consultancy and interaction between physical space and business performance, (3) Branding and communication design, (4) Data center design, and

(5) Hospitality and residential luxury design. Additionally, firm X is organized into three business units (BU):

1. Urban and Building, specialized in the architectural design of buildings for residential, commercial and cultural uses;
2. Engineering and Sustainability, structured to manage large mechanical and electrical (M&E) design projects, with a focus on sustainability and energy saving;
3. Retail.

The close collaboration between this study and firm X was carried out in partnership with the 'Urban and Building' and the 'Engineering and Sustainability' business units in particular.

Traditionally, firm X has implemented a design process enabled by CAD technologies. Procedures and policies of this traditional design process are standardized and certified in the firm's Quality Management System (QMS).

At the time of the firm's collaboration with this study (2014–2016), there was little understanding and awareness of the changes associated with digital transformation. Some senior managers had a visionary understanding and pushed the firm to shift toward a collaborative and iterative digital design process. The objective of this shift was three-fold: (1) to improve internal capacity and efficiency, (2) to attain competitive advantage and a better position in the market place, and (3) to provide sustainable and optimized design solutions.

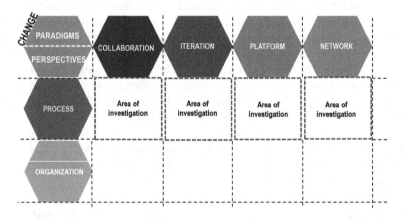

Fig. 5.1 Primary areas of investigation

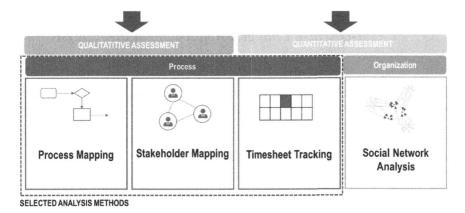

Fig. 5.2 Process mapping, stakeholder mapping, timesheet tracking selected to assess the process-oriented changes in firm X

5.2 Identifying, Framing, and Assessing Process-Oriented Changes: Application of the Methodology

The methodology for change management proposed was applied to firm X in order to identify, frame, and assess the process-oriented changes associated with the firm's ongoing digital transformation. First, the *conceptual framework* proposed was used to identify the primary areas of transformation (see Fig. 5.1). Second, a selection from the *set of analysis methods* proposed was made to perform qualitative and quantitative assessments of process-oriented changes (see Fig. 5.2). Lastly, information were extracted from the performed business analyses to draw and calibrate accordingly the strategy for transitioning toward a collaborative and iterative digital design process.

5.2.1 Identifying Areas of Transformation and Framing the Associated Changes: Application of the Conceptual Framework

The conceptual framework proposed by this study was applied and tested on firm X in order to identify the areas of transformation associated with the firm's transition to a collaborative and iterative digital design process. Consequently, the process-based aspects of the framework proposed were identified as primary investigation areas for firm X (see Fig. 5.1). Additionally, due to the importance of promoting a platform-based and networked project organization for the management of the project stakeholders involved in a collaborative and iterative digital design process, the related investigation areas were also identified as fundamental elements to be considered for the firm's digital transformation.

5.2.2 Identifying, Visualizing, and Assessing Process-Oriented Changes: Application of the Analysis Methods

The set of analysis methods was applied to assess the process-oriented changes associated with the firm's objective of transitioning toward a collaborative and iterative digital design process. Specifically, process mapping, stakeholder mapping, and timesheet tracking were selected to perform both qualitative and quantitative assessments of change (see Fig. 5.2).

5.2.3 As-Is, Transition, and To-Be: Levels of Analysis of Process-Oriented Changes

Assessments through process mapping, stakeholder mapping, and timesheet tracking were performed at the following three levels:

1. 'As-is' state, which refers to the design process traditionally implemented by the firm;
2. 'Transition' state, alias the process implemented to develop BIM pilot projects. To achieve this, a BIM pilot project being developed by the firm was selected as case-study;
3. 'To-be' state, which is the envisioned digital design process.

These three levels of analysis were associated with three different colors in the development of three process maps: grey to indicate tasks of the traditional design process (see Fig. 5.3), orange to draw activities performed in a BIM pilot project (see Fig. 5.4), and blue to color the tasks of the envisioned digital design process (see Fig. 5.5).

As-Is

First, data about the traditional design process were collected. The firm's Quality Management System (QMS) was analyzed to understand the procedures of the traditional design process. The main objective of this analysis was to later identify changes in the traditional design process, that means activities either added, modified, or eliminated by comparison with both the digital pilot process ('transition' assessment level) and the envisioned digital design process ('to-be' assessment level).

Transition

Second, a BIM pilot project was selected as a case-study, the related project development process was mapped (qualitative assessment), and the project timesheet tracked and analyzed (quantitative assessment). The goal of this analysis was to understand both opportunities and challenges of the firm's digital transformation as currently managed. To achieve this, a comparison with the envisioned digital design process

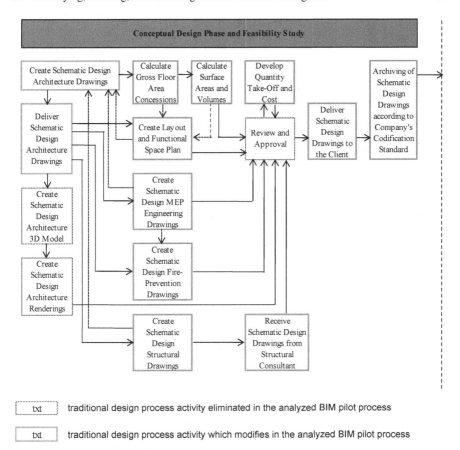

Fig. 5.3 Extract from the traditional design process map: focus on the phase of conceptual design and feasibility study

('to-be' assessment level) was performed. In this way, it was possible to target the ongoing transition to a collaborative and iterative digital design process according to the benchmark of the envisioned digital design process.

To-Be

Lastly, international standards were reviewed to understand how to set the benchmark of the envisioned digital design process by reference to which calibrating the firm's digital transformation strategy.

To achieve this, the following were the main steps performed in the application of the methodology proposed in Firm X:

1. Reviewing the firm's Quality Management System;
2. Breaking down the traditional design process in phases and related activities;

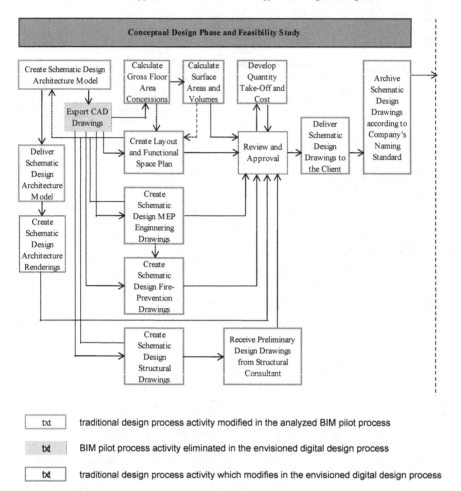

Fig. 5.4 Extract from the process map of the BIM pilot project: focus on the phase of conceptual design and feasibility study

3. Defining outputs (e.g. drawings, documents, etc.) and involved stakeholders for each activity;
4. Drawing the process map of the traditional design process (see Fig. 5.3);
5. Selecting a BIM pilot project as case-study;
6. Drawing the process map of the BIM pilot project development process (see Fig. 5.4);
7. Mapping the project stakeholders involved in the digital pilot project;
8. Tracking the timesheet of the BIM pilot project;
9. Analyzing the timesheet in terms of positive and negative aspects of the transition to a collaborative and iterative digital design process (i.e. time delay due to the lack of project stakeholders' digital skills and capabilities);

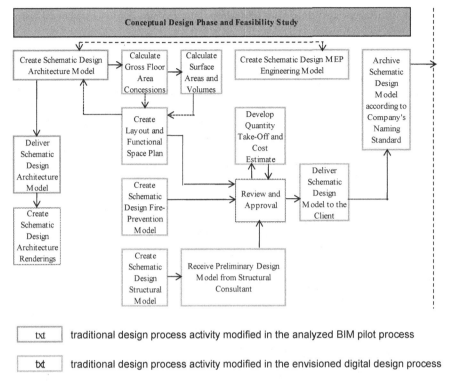

txt   traditional design process activity modified in the analyzed BIM pilot process

txt   traditional design process activity modified in the envisioned digital design process

Fig. 5.5 Extract from the process map of the envisioned digital design process: focus on the phase of conceptual design and feasibility study

10. Identifying changes in the traditional design process by comparison with the BIM pilot project development process;
11. Reviewing international digital and BIM standards;
12. Developing the process map of the envisioned digital design process according to the reviewed standards (see Fig. 5.5);
13. Identifying changes, in other words, activities to be added, modified and eliminated, in the BIM pilot process by comparison with the envisioned digital design process map;
14. Drawing and calibrating the firm's digital transformation strategy according to both the qualitative (process maps and stakeholder maps) and the quantitative (project timesheet) analyses.

Table 5.1 Differences highlighted in the BIM pilot process in comparison to the traditional design process

Process	People	Product	Policy
Integrated digital workflow between the architecture and engineering areas	Digital and BIM training offered to the architecture and engineering areas	Reduced file exchange between the architecture and rendering areas	Enlarged marketing opportunities
Integrated digital workflow between the architecture and rendering areas	Improved synergy and collaboration between the architectural and engineering teams	Improved consistency and accuracy of the architectural and engineering project outputs	Hardware and software implementation
	New roles and responsibilities within the project team	Reduced file production by the architecture and engineering teams	R&D investment
	Improved design consciousness of the architectural and engineering areas		

5.2.4 Evaluating Process-Oriented Changes: Challenges and Opportunities

After having performed qualitative (process mapping and stakeholder mapping) and quantitative (timesheet tracking) assessments, a phase of interpretation of the business analyses was performed to evaluate the process-oriented changes identified. First, starting from the three process maps developed, a synthesis about the process-oriented changes identified is proposed through the development of summary tables (see Tables 5.1 and 5.2). These tables highlight process activities either added, modified, or eliminated in the traditional design process by comparison with the analyzed BIM pilot project. Second, starting from the stakeholder mapping performed, this study analyzed the relationships between project stakeholders in order to understand whether it was a fully-digital (BIM to BIM) or mixed information exchange (BIM to CAD and vice versa). Lastly, starting from the timesheet tracking of the BIM pilot project analyzed, an analysis of positive and negative aspects of the BIM pilot project was performed. Specifically, it was assessed the time (man-hours) spent on activities either boosting or slowing down the transition to a collaborative and iterative digital design process.

Process mapping, carried out at the three levels of analysis, helped to identify what changes were sought in the traditional design process by comparison, first, with the BIM pilot process, and then with the envisioned digital design process. This analysis was useful to understand the following aspects:

Table 5.2 Differences highlighted in the BIM pilot process by comparison with the envisioned digital design process

Process	People	Product	Policy
BIM not implemented throughout all the phases of the process	BIM training not offered to all the internal departments involved in the pilot project	Lacking preliminary standardization of the BIM environment according to business processes and procedures	Lacking preliminary agreement with the Client about management, technical and legal of the BIM process by defining EIR—Employer's Information Requirements and developing a BEP—BIM Execution Plan
BIM not implemented in all the working departments involved in the BIM pilot project	Lacking preliminary definition of project stakeholders' roles and responsibilities		
	Lacking preliminary evaluation of project consultants' BIM skillset		

– What activities traditionally performed in the design process were changed;
– How the sequence of activities changed;
– Which actors were affected by the change and how the related roles and responsibilities might change accordingly;
– How information exchange between project parties modified in a digital environment.

A summary table highlights the differences in the traditional design process through comparison with the pilot digital process (see Table 5.1). Another table outlines the differences in the BIM pilot process through comparison with the envisioned digital design process (see Table 5.2).

5.2.5 Identifying Barriers to the Transition to a Collaborative and Iterative Digital Design Process

A second level of interpretation was carried out starting from the stakeholder mapping performed. Specifically, this study analyzed the relationships between project stakeholders in order to understand whether it was a fully-digital (BIM to BIM)

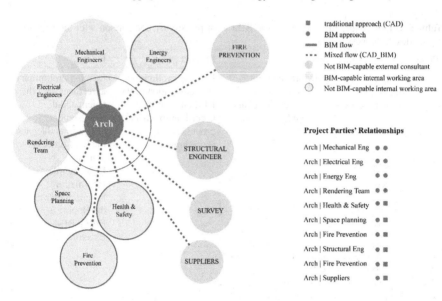

Fig. 5.6 Fully digital (blue lines) and mixed (greyed dotted lines) workflows between project parties

or mixed information exchange (BIM to CAD and vice versa) (see Fig. 5.6). As described in Fig. 5.6, the following were the mixed flows (dotted line):

- Architects—Fire Prevention team (external consultant);
- Architects—Structural engineer (external consultant);
- Architects—Survey team (external consultant);
- Architects—Suppliers (external consultant);
- Architects—Energy engineers (internal);
- Architects—Health and Safety team (internal);
- Architects—Fire prevention team (internal);
- Architects—Space planning team (internal).

Lastly, starting from the timesheet tracking performed, an analysis of positive and negative aspects of the BIM pilot project was carried out. Specifically, it was assessed the time (man-hours) spent on activities either boosting or slowing down the transition to a collaborative and iterative digital design process. The BIM pilot project, in fact, recorded a delay time which was around 20% of the initial estimated time. Therefore, this study tried to understand which issues determined this time overrun (see Figs. 5.7 and 5.8).

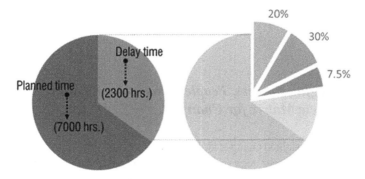

Fig. 5.7 Time analysis of the digital pilot project. In dark-grey, the planned time, in light-grey the delay time

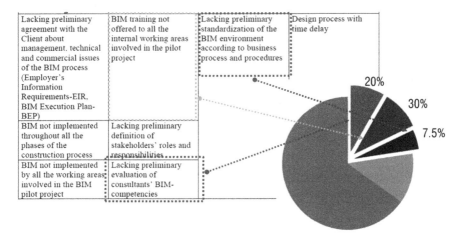

Fig. 5.8 Problematic issues determining a delay time of the digital pilot project

5.3   Drawing and Calibrating the Strategy for Transitioning toward a Collaborative and Iterative Digital Design Process

The aforementioned analysis phase was followed by a collective postulation of possible change management strategies, from which a single plan of action emerged. The developed change management strategy was targeted according to the valuable information collected through the performed qualitative and quantitative process-based analyses. The proposed change management strategy focused on the management of the following interpretation lenses of change:

– Process;
– People;

– Product;
– Policy.

5.3.1 Managing Process, People, Product, and Policy: Strategic Matrix for Change

Given the process-oriented changes identified, a potential digital transformation strategy was developed. The strategy proposed took into consideration the 4Ps aspects of change—process, people, product, and policy—and associated each of them to the achievement of strategic targets. In particular:

Process

– Defining new roles and responsibilities;
– Involving all the business units in the firm's digital transformation;
– Revising the firm's QMS according to a collaborative and iterative digital design process;
– Turning temporary and adversarial relationships into collaborative and long-term partnerships;
– Creating a network of trusted and digitally-capable consultants.

People

– Training all the firm's business units;
– Training all the firm's levels of authority (both working teams and project managers);
– Improving competencies of digitally-capable personnel;
– Spreading innovation consciousness of the firm's digital transformation through workshops and events.

Product

– Standardizing the digitally-enabled work environment by developing business templates and BIM objects' warehouse;
– Evaluating the purchase of interoperable digital software and tools for different business units.

Policy

– Developing or adopting standards and procedures for data exchange, deliverables, technology infrastructure, and legal framework.

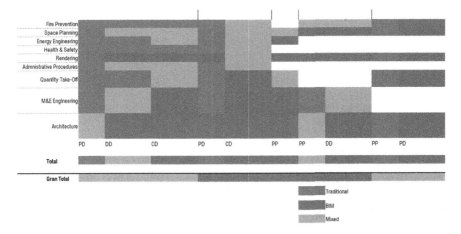

Fig. 5.9 Heat map to monitor the firm's transition to a collaborative and iterative digital design process

5.3.2 Monitoring the Transition to a Digital Design Process: Heat Map of Digital Projects

Lastly, this study developed a heat map to visualize the digital maturity of the diverse projects developed by the firm after the first BIM pilot project analyzed in this study (see Fig. 5.9). Each of the projects monitored was described by the process phases implemented and the business units involved. Additionally, for each of them, it was indicated whether it was managed with a traditional (yellow boxes), 'mixed' (grey boxes), or fully digital (blue boxes) approach. In this way, for example, it was possible to understand which business units were more advanced or lacking in terms of digital skills and capabilities and to target the change management accordingly.

Chapter 6
Application of the Methodology for Change Management: The Case of a Canadian Multidisciplinary Design Firm

Abstract The aim of this chapter is to present the application and testing of the methodology for change management in a Canadian multidisciplinary design firm. The objective of this firm was to move toward a platform-based and networked organizational structure. First, the conceptual framework proposed is used to identify the areas of transformation associated with the firm's strategic objective. Second, the set of change analysis methods is employed to understand the existing organizational structure, what organizational changes occur because of the firm's ongoing digital transformation, and to envision the future platform-based and networked organizational structure. To achieve this, Social Network Analysis (SNA) is the analysis method applied in this case study. This method aims at identifying, visualizing, and assessing organizational changes at the following three levels:

- 'As-is' state, which refers to the existing vertical and hierarchical organizational structure;
- 'Transition' state, alias the emergent network for digital support and advice;
- 'To-be' state, which is the envisioned platform-based and networked organizational structure.

Lastly, a potential digital transformation strategy is developed in accordance with the information extracted from the SNA. The strategy proposed translates into two actions for organizational change: (1) Definition of a new organizational service in the form of a shared group for digital support and advice, and (2) Reconfiguration of the existing organizational structure according to a platform-based and networked model.

Keywords Digital transformation · Multidisciplinary design firms · Organizational change · Platform · Network · Social Network Analysis (SNA)

6.1 Transitioning to a Platform-Based and Networked Organization: The Firm's Objective

Firm Y is one of Canada's leading architecture and engineering groups. The firm is organized into the following business operating units (BOU):

© The Author(s), under exclusive license to Springer Nature Switzerland AG 2019   65
M. M. Bonanomi, *Digital Transformation of Multidisciplinary Design Firms*,
PoliMI SpringerBriefs, https://doi.org/10.1007/978-3-030-19701-8_6

- Buildings;
- Environmental Services;
- Energy and Resources;
- Water;
- Infrastructure.

The close collaboration between this study and firm Y was focused on knowledge creation and sharing and specifically how advice and support-seeking relationships and networks in organizations could provide a competitive advantage in knowledge-intensive work.

At the time of the firm's collaboration with this study (2017), there was little understanding and awareness of the organizational changes associated with the firm's ongoing digital transformation. Some senior managers had a visionary understanding and pushed the firm to shift toward a more platform-based and networked organizational structure. The objective of this shift was three-fold: (1) To improve internal process efficiency and people performance, (2) To retain competitive personnel, and therefore a better position on the market, and (3) To develop a new organizational service in the form of a shared group for digital support and advice.

6.2   Identifying, Framing, and Assessing Organizational Changes: Application of the Methodology

The methodology for change management proposed was applied and tested on firm Y to identify, frame, and assess the organizational changes occurring because of the firm's ongoing digital transformation. First, the *conceptual framework* proposed was used to identify the primary areas of transformation associated with the firm's strategic objective to move toward a more platform-based and networked organizational structure. Second, a selection from the *set of analysis methods* proposed was made to perform a quantitative assessment of organizational changes. Lastly, information were extracted from the performed business analyses to draw and calibrate accordingly the transition toward a platform-based and networked organizational structure.

6.2.1   Identifying Areas of Transformation and Framing the Associated Changes: Application of the Conceptual Framework

The conceptual framework proposed by this study was applied and tested on firm Y in order to identify the primary transformation areas associated with the firm's transition toward a platform-based and networked organizational structure (see Fig. 6.1).

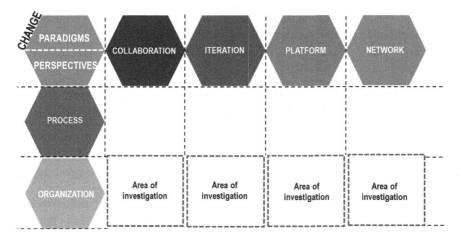

Fig. 6.1 Primary areas of investigation for the digital transformation of firm Y

Consequently, the organizational aspects of the framework proposed were identified as primary investigation areas for firm Y.

6.2.2 Identifying, Visualizing, and Assessing Organizational Changes: Application of the Analysis Methods

The set of analysis methods was applied to assess the organizational changes associated with the firm's transition to a platform-based and networked organizational

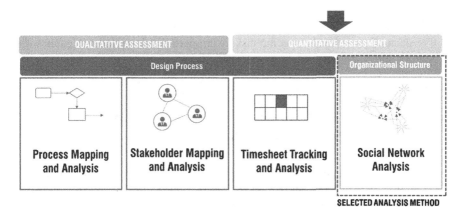

Fig. 6.2 Social network analysis selected to assess the process-oriented changes brought on by the digital transformation of firm Y

structure. Specifically, social network analysis was selected to perform a qualitative assessment of organizational changes (see Fig. 6.2).

6.2.3   As-Is, Transition, and To-Be: Levels of Analysis of Organizational Changes

Assessments through SNA were performed at the following three levels of analysis:

1. 'As-is' state, which refers to the existing vertical and hierarchical organizational structure;
2. 'Transition' state, alias the emergent network for digital support and advice;
3. 'To-be' state, which is the envisioned platform-based and networked organizational structure.

SNA was aimed at specifically answering the following research questions:

- Who are the emergent informal connectors for digital support and advice?
- Where are they located within the existing formal hierarchy? Are they peripheral specialists or are they well-spread across the different hierarchical levels?
- Which are their existing formal reporting relationships? Are they efficient and effective in terms of digital knowledge creation and sharing?
- How are the emergent informal connections for digital support and advice configured?
- What about bringing together these emergent informal connectors for digital support and advice into a shared 'Support Digital Group'?

Starting from these questions, two distinct, but interrelated, levels of analysis were completed:

- Analysis of the existing formal hierarchy;
- Analysis of the emergent informal network.

For each of these levels, specific tasks were performed, related outcomes were produced, and data-sources defined (see Table 6.1).

6.2.4   Assessing the Existing Formal Hierarchy: As-Is State of the Organizational Structure

Assessing all the organizational relationships in play within firm Y led to identify the following two different types of relationships: (1) Reporting formal relationships as defined by the existing organizational structure ('as-is' state), and (2) Advice and support-seeking connections creating an emergent network for digital knowledge ('transition' state).

Table 6.1 Task, outcome, and data-source for the analyses of both the existing formal hierarchy and the emergent informal network for digital support and advice

	Task	Outcome	Data source
Analysis of the existing formal hierarchy	Mapping the existing reporting relationships	Hierarchical cluster and radial map visualizing the existing reporting relationships (at the firm level)	HR database specifying all the firm's employees and the related supervisors
	Visualizing the formally-designated organizational key-players: connectors and brokers	Network map visualizing the formally-designated connectors (at the BOU level) Network map visualizing the formally designated brokers (at the BOU level)	HR Database specifying all the firm's employees and the related supervisors
Analysis of the emergent informal network for digital support and advice	Identifying the key-players of the emergent informal network	Network map visualizing the key-players of the emergent informal network and the related formal reporting relationships	Database of BIM/CAD leaders available from the firm's website HR Database specifying all the firm's employees and the related supervisors
	Mapping their formal reporting relationships and highlighting any related inefficiency and underutilized potential		
	Mapping the advice and support-seeking connections defining the emergent informal network	Network map visualizing connections and connectors of the emergent informal network	Survey database about the emergent informal connections and connectors for digital knowledge

The formal organizational structure was assessed through the development of a hierarchical cluster map. Formally-designated connectors[1] (see Fig. 6.3) and brokers[2] (see Fig. 6.4), as well as the related leadership structure, were visualized through network maps.

---

[1]Connectors can be defined as go-to people. They serve the firm in which they work by linking colleagues in hubs. They are therefore located at the center of subnetworks.

[2]Brokers are those people who connect various subnetworks together. They can connect either different BOUs (information brokers) or office locations (geographical brokers). They serve the firm in which they work by breaking down the silos that may obstruct collaboration and knowledge-sharing. They are therefore located in the "white space" of network maps.

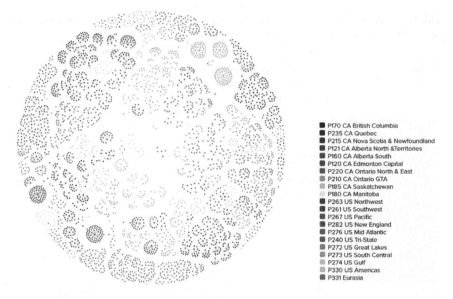

Fig. 6.3 Map of the existing formal structure of the 'Buildings' BOU. Formally-designated connectors are located at the centre of each subnetwork

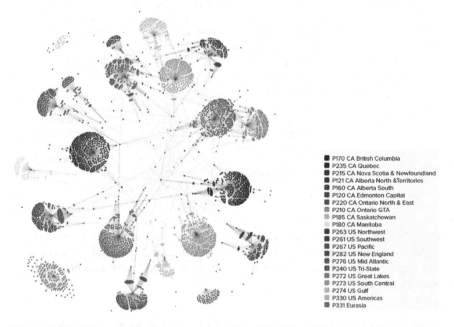

Fig. 6.4 Map of the existing formal structure of the 'Buildings' BOU clustered by regions. Formally-designated (geographical) brokers are in the white space between each regional hub

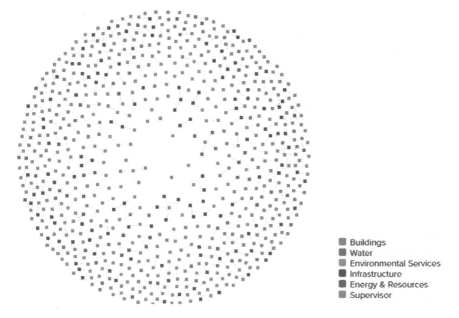

Fig. 6.5 Network map visualizing the formal reporting relationships of the emergent informal go-to people for digital knowledge. Informal go-to people for digital knowledge are coloured by BOU—in grey the related supervisors

When this study focused on the emergent informal connectors for digital knowledge, the visualization of their formal reporting relationships had nothing in common with a network. It looked like a quite sparse and fragmented structure made up of multiple sets of relationships (see Fig. 6.5). This was a problem because this organizational pattern did not help that personnel, who was performing analogous tasks, to connect one to each other. In other words, employees, who were doing similar work, were not networked but rather isolated in multiple sets of relationships (see Fig. 6.5). What about the efficiency of these formal relationships in terms of digital knowledge creation and sharing?

A typical common goal of reorganizations is bringing together employees who perform related tasks but are dispersed in different businesses or locations. Therefore, this study aimed at identifying emergent informal connectors for digital knowledge, alias the go-to people across the firm who developed, deployed, and supported the technologies used to deliver work to clients, and at bringing them together for the sake of pushing the boundaries on firm's digitalization.

6.2.5 Mapping the Emergent Informal Network for Digital Support: Transition State of the Organizational Structure

Krackhardt and Stern from Cornell University (1988) argued that "emergent networks, left to themselves without the aid of conscious design, will form naturally in ways that are suboptimal, even dysfunctional, for the organization". They also argued that "the degree to which the informal organization is designed optimally is measurable."

Given these considerations, social network analysis should play a crucial role in drawing and calibrating organizational change management. SNA can help to make emergent networks more transparent, and therefore revealing opportunities for either organizational change or integration. Specifically, SNA can be used to identify emergent and informal organizational key-players, to highlight well-functioning connections and economies of scale, which a firm should avoid disrupting and to optimally design emergent and informal networks, not formally engineered nor configured yet.

When this study analyzed the existing formal reporting relationships of the emergent informal connectors for digital support and advice, it found out that they were configured into a quite dispersed structure (see Fig. 6.5). This structure was defined by a sparse and fragmented set of dyadic formal relationships. This was a problem because employees performing similar tasks with analogous skills and expertise were not connected, but rather completely isolated one from each other. Therefore, bringing together these digital experts through a shared Application Support Group would let the organization gain benefits from multiple perspectives, namely knowledge management, business process efficiency, and employees' performance and satisfaction.

6.2.6 Envisioning a Digital Support Group: To-be State of the Organizational Structure

Many studies about networks and knowledge transfer show how organizational structures might affect the diffusion of information (Allen 1977, Burt 1992, Rogers 1995). Furthermore, many studies argue that interpersonal relationships within organizations are central to finding information and generating actionable knowledge (Allen 1977, Rogers 1995). Many studies have shown also that relationships in organizational settings have a dyadic nature: "They are established in part through relative position in formal structure (Lincoln 1982) and in part through repeated interaction and social psychological processes (Weick 1979)" (Cross and Sproull 2004). If these relationships are not aligned one to each other, process efficiency might be hindered by formal reporting relationships, which sometimes obstacle people to reach the knowledge they are seeking. Consequently, because of a lacking knowledge distri-

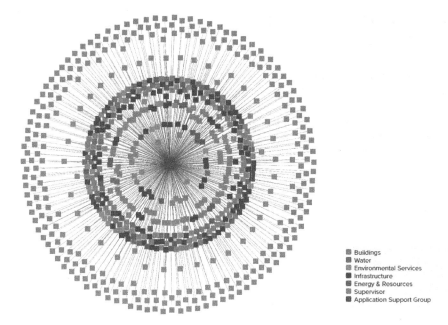

Fig. 6.6 Network map visualizing the Digital Support Group proposed

bution network, employees typically start connecting informally by reaching out to each other and creating new work relationships.

Given these considerations, it is of great importance to manage both the formal and informal organizational relationships. Efficient and effective knowledge creation and sharing, in fact, occurs when both formal reporting and informal supporting relationships are recognized, managed, and aligned one to each other.

After SNA was performed within firm Y, the need for an organizational struc-ture reconfiguration was demonstrated. To achieve this, it was decided to formally empower the emergent informal network for digital support by bringing together people who were doing similar work, but they were not connected one to each other, in order to achieve economies of scale while sharing best practices and expertise across the organization. The creation of an Digital Support Group could also help firm Y in improving business process efficiency by providing an efficient workflow letting digital knowledge reach the people best able to exploit it (see Fig. 6.6). In this sense, boosting connectivity between go-to people for digital support and advice through the creation of a shared Digital Support Group reduced the number of steps required for any individual in the network to get in touch with a colleague, as well as increased the connectivity of the peripheral people.

6.3   Evaluating Organizational Changes: Challenges and Opportunities

There are multiple risks associated with overlooking the emergent network for digital knowledge. These risks can be associated with the following categories:

- Knowledge management;
- Business process efficiency;
- Employees' performance and satisfaction.

6.3.1   Knowledge Management

"If we think networks are important for performance and sharing best practices, then executives need to know who the key players are and whether they are in a role that is enabling others or potentially constraining others by hoarding information." (Cross et al. 2013).

Digital experts, who are usually well connected to knowledge sources outside the firm (users' groups, blogs, conferences, university labs, etc.) must possess strong connections internally as well. If digital processes occur within a fragmented structure, rather than within a networked community of experts, there is the risk of losing opportunities for knowledge creation and innovation outcomes. Without an effective knowledge distribution network, the contributions of firm's experts to digital innovation are limited.

6.3.2   Business Process Efficiency

"It has become a truism to say that networks are important for success" (Cross et al. 2013).

The other side of just having formal network centrality is the danger of overload. Formally-designated connectors, alias firm's leaders, who are not able to let go, might invisibly and unintentionally slow others down. Furthermore, as already mentioned before, digital experts are often reporting to leadership that is not familiar with their skillset, but this should not turn out to be a risk in terms of business process efficiency. That can be an opportunity to look at role definitions and other things to alleviate central overload points.

6.3.3   *Employees' Performance and Satisfaction*

"People who are not getting formal recognition tend to be the biggest flight risks. Looking at centrality gives executives the opportunity to pay attention to hidden stars whom they had not been thinking about before." (Cross et al. 2013).

There is a disproportionate influence of certain people in the center of the network. As Cross et al. (2013) underline, "What I tend to find is that of the people who are really enabling their colleagues to be successful—in other words, helping others to do good work—about 50% of them just haven't been recognized. They are not in the top talent list; they are not known by the leaders or on their radar screen." This represents a potential flight risk because digital experts who are unsatisfied of not getting the appropriate recognition for the work done are most likely prepared to accept a working proposal from another firm. In support of this statement, it should be noted that Firm Y has been struggling to keep digital experts' staff.

References

Allen T (1977) Managing the flow of technology. MIT Press, Cambridge, MA

Burt R (1992) Structural holes. Harvard University Press, Cambridge, MA

Chinowsky P, Taylor J (2012) Networks in engineering: an emerging approach to project organization studies. Eng Proj Organ J 2(1–2):15–26

Chinowsky P, Diekmann J, O'brien J (2010) Project organizations as social networks. J Constr Eng Manag 136(4):452–458

Cross R, Sproull L (2004) More than an answer: information relationships for actionable knowledge. Organ Sci 15(4):446–462

Cross R, Kaše R, Kilduff M, King Z, Minbaeva D (2013) Bridging the gap between research and practice in organizational network analysis: a conversation between Rob Cross and Martin Kilduff. Human Res Manag 52(4):627–644

Krackhardt D, Stern RN (1988) Informal networks and organizational crises: an experimental simulation. Social Psychol Q 51(2):123–140

Lincoln J (1982) Intra- (and inter-) organizational networks. Res Sociol Organ 1:1–38

Rogers EM (1995) Diffusion of innovations. Free Press, New York

Weick KE (1979) The social psychology of organizing. McGraw-Hill, New York

Conclusions

This study investigates the changes associated with the digital transformation of multidisciplinary design firms from the perspectives of process and organization. The main objective of this study is to bring forth a better understanding of what changes occur in the design process and organizational structures of multidisciplinary design firms in periods of digital transformation, how they unfold in practice, and what change management strategies can be developed to plan and lead the digital transformation accordingly.

Although the significant influence of the context of change on digital transformation, multidisciplinary design firms do not reveal to be fully aware of this. The context of change seems to be implicitly considered. Frameworks and methods for management of change related to digital transformation are often quite 'general' and not based on or informed with data about the specificity of the firm and reference market.

To the contrary, the definition of a successful digital transformation depends importantly on the specific context of change, that includes, at least: the organizational model of the firm and its level of integration; client's goals and organization; readiness of reference market; and stakeholders' digital maturity across the supply chain. Being aware of the influencing factors coming from the specific context of change means to be able to lead a risk-informed digital transformation.

Additionally, design firms should keep in mind that nowadays a helpful set of international standards is available (e.g. guidelines, criteria, framework of tools, etc.) regarding most of the elements involved in digital transformation, such as standards for classification and coding, regulatory framework, rules for interoperability, etc. Being aware of these issues means for digitally-evolving design firms to acquire criteria, methods, and tools to manage the process-oriented and organizational changes associated with digital transformation in context and both at strategic and operational levels.

Some keywords and statements could be considered for implementing a systematic approach to change management associated with digital transformation:

- *Knowledge for Change*

Knowledge must be the core and value of change management. Combining knowledge and change management means setting the ground for collecting data for change and referring them to the specific context and objectives of change. Decisions and operations associated with digital transformation depend significantly on knowledge: the more the knowledge for change expands and the more the abilities of planning and leading change increase. Knowledge for change may concern many subjects, such as the organizational model of the firm and its level of integration; client's goals and organization; readiness of market of reference; digital (im-)maturity of stakeholders across the supply chain, etc.

- *Framing Change*

The condition for collecting data is to frame change into a 'breakdown structure'. This means assuming a framework, like the one proposed by this study, to identify paradigms, perspectives of change, and context of change. Many international standards may be assumed as references to organize this framework. This framework should allow to implement gradual change management in the specific context of change and at various stages of implementation over time.

- *Gradualism in Change*

All the aspects connected to change must be considered according to a criterion of gradualism. Strategies for change must grow over time through a continuous data collection about process and practices. This means doing measurements and benchmarks, collecting data coming from various project processes and gathering feedback information from various project parties. Change management strategy must follow digital transformation throughout its whole duration, entrusting its implementation to various stakeholders, who will change across the different stages of transition.

- *Analysis for Targeting Change*

A systematic methodology, like the one proposed by this study, can be considered as a supporting tool for collecting data about change and making them available both for decision making and operational tasks. Whatever is the strategy, an analysis-based approach must be defined preliminarily. Furthermore, the implementation process of an analysis-based system must follow the criterion of gradualism, growing over time in accordance with digital maturity levels of the firm and reference market.

In conclusion:

I. Digital transformation is becoming increasingly urgent for design firms: open problems of efficiency, productivity, and competitiveness in the AEC industry have demonstrated significant weaknesses associated with the traditional design process and organizational models. To move toward new forms of process and organization, institutional players, technical-scientific community, researchers and practitioners should devote increasing attention to this topic;

II. In the last decades, digitally-mature design firms started undergoing paradigm shifts at process, organization and technological levels with remarkable results. However, due to various risks associated with these changes, a wide-spread diffusion and application of new forms of process and organization is still lagging behind;

III. Potentials and challenges of digital transformation have been widely investigated by national and international researchers and practitioners acting as first movers of this topic. However, a thorough and systematic analysis of the related process-oriented and organizational changes in practice is needed to manage successfully both changes and risks associated;

IV. Since a strategy for change is the primary element to be set up for a successful digital transformation, a conceptual framework and a set of change analysis methods are herewith proposed. Since the design process and the organizational structures are being widely affected, the present study suggests the adoption of a double lens of perspective: process and organization;

V. On the process side, interpretative process mapping and comparison of the design process at varying levels (traditional - pilot - envisioned) allows understanding what changes occur and how they should be managed in practice. The goal is to target the shift toward a collaborative and iterative digital design process. On the organizational side, social network analysis helps to observe actual working relationships among firm's employees to ascertain who is working with whom in periods of digital transformation and therefore revealing an opportunity for either organizational structure change or integration based on the new roles and relationships brought on by digital transformation.

Printed in the United States
By Bookmasters